DIAMOND OPTICS

Volume 969

CONTENTS

(continued)

DIAMOND OPTICS

Volume 969

CONFERENCE COMMITTEE

Chairs
Albert Feldman
National Institute of Standards and Technology

Sandor Holly
Rockwell International Corporation

Cochairs
Andrzej R. Badzian, Pennsylvania State University
Robert Nemanich, North Carolina State University
K. V. Ravi, Crystallume
Sidney Singer, Los Alamos National Laboratory
John A. Woollam, University of Nebraska/Lincoln

Session Chairs
Session 1—Deposition and Processing I
Albert Feldman, National Institute of Standards and Technology

Session 2—Deposition and Processing II
Andrzej R. Badzian, Pennsylvania State University

Session 3—Characterization I
John A. Woollam, University of Nebraska/Lincoln

Session 4—Characterization II
W.-K. Chu, University of North Carolina

Session 5—Diamond Applications
Sidney Singer, Los Alamos National Laboratory

Session 6—Diamond Optical Applications
K. V. Ravi, Crystallume

Conference 969, *Diamond Optics*, was part of a four-conference program on Optical Materials held at SPIE's 32nd Annual International Technical Symposium on Optical & Optoelectronic Applied Science and Engineering, 14–19 August 1988, San Diego, California. The other conferences were

Conference 968, *Ceramics and Inorganic Crystals for Optics, Electro-Optics, and Nonlinear Conversion*
Conference 970, *Properties and Characteristics of Optical Glass*
Conference 971, *Nonlinear Optical Properties of Organic Materials.*

Program Chair: **Solomon Musikant,** General Electric Company

DIAMOND OPTICS

Volume 969

INTRODUCTION

Recently, deposition processes have been discovered that have resulted in the synthesis of diamond films at significantly lower pressures and temperatures than previous methods of diamond synthesis. This new technology, together with the earlier methods for growing diamonds and diamond-like carbon, offer the promise of superior optical components because of the unique properties of diamond. Crystalline diamond is both the hardest material known and the material with the largest thermal conductivity at room temperature. In addition, it is transparent over large spectral ranges, it is chemically inert, it is highly impervious, and it is stable at high temperatures.

In the context of this proceedings, we treat diamond and diamond-like carbon on an equivalent basis, even though there are many differences between the two types of materials. For example, diamond-like carbon films are significantly smoother than diamond films, but they have lower thermal conductivities. Diamond-like carbon actually represents a range of hard materials based on carbon and usually containing large fractions of hydrogen. In terms of chemical bonding, diamond is composed of sp^3 bonds exclusively, whereas diamond-like carbon is composed of a mixture of sp^2 and sp^3 bonds. The invited paper by J. C. Angus discusses these concepts in great depth.

The use of diamond as an optical material will have many significant ramifications. At present, many optical materials are subject to degradation due to environmental factors, including, chemical attack from corrosive atmospheres as in excimer lasers, or abrasion and erosion damage from atmospheric conditions such as wind, rain, and dust impact. Diamond offers the possibility of a superior protective optical coating material that is transparent over an extensive range of the spectrum. In addition, the high thermal conductivity of diamond would permit the fabrication of optics resistant to thermal distortion and degradation caused by the absorption of intense optical radiation. For example, the use of diamond as the output window would permit extraction of significantly greater power from and/or more compact construction of free-electron lasers.

Several problems must be solved before diamond can be used extensively as an optical material. Currently, the diamond films produced exhibit a significant surface roughness and optical scatter. Furthermore, as a coating material, the deposition temperatures are higher than the softening points of many optical substrate materials. The small size of single crystal diamond presently limits the size of diamond windows, and the high cost of these windows would limit their use to only specialized applications. Diamond-like carbon films represent a mature field and are currently used commercially as antireflecting and protective coatings on Si and Ge optics for the infrared.

DIAMOND OPTICS

Volume 969

The purpose of this proceedings is to present to the optical engineering community the most recent developments in diamond and diamond-like carbon technology for optical applications. The proceedings is based on the organization of the conference, which consisted of six sessions divided into four principal areas: Deposition and Processing; Characterizations; Diamond Applications; and Diamond Optical Applications. By reading this proceedings we are certain that the reader will be impressed by the significant strides being made and the great excitement being generated by this new technology.

We would like to thank all of the participants who have made this conference a great success: the cochairs, the session chairs, the authors, and SPIE.

Albert Feldman
National Institute of Standards and Technology
(Formerly National Bureau of Standards)

Sandor Holly
Rockwell International Corporation
Rocketdyne Division

SESSION 1

Deposition and Processing I

Chair
Albert Feldman
National Institute of Standards and Technology

Diamond and "diamondlike" phases grown at low pressure: growth, properties and optical applications

John C. Angus
Cliff C. Hayman
Richard W. Hoffman

Case Western Reserve University
Cleveland, Ohio 44106

ABSTRACT

A wide variety of energetically assisted methods have been employed to grow diamond films at low pressures. Common features of the processes include the presence of atomic hydrogen, energetic carbon containing fragments and high surface mobilities. Some understanding of the molecular processes taking place during nucleation and growth of diamond has been achieved, but detailed molecular mechanisms are not known with certainty. Application of vapor grown diamond for abrasive grit, tool coatings and wear resistant surfaces can be expected shortly. However, the use of vapor grown, crystalline diamond in optical applications or as active semiconductor elements will require further control over surface roughness and crystalline quality.

Related research has led to the discovery of a new class of materials, the so—called "diamondlike" phases. Two types of diamondlike materials may be distinguished, namely, the diamondlike hydrocarbons and the diamondlike carbons. These materials possess exceptional hardness, smoothness and chemical inertness. They show promise as combined anti—reflection and abrasion resistant coatings on optical elements, as protective coatings on magnetic and optical disks, as diffusion barriers and for photo—lithographic applications.

1. INTRODUCTION

It has been known for over twenty years that crystalline diamond could be grown by chemical vapor deposition from hydrocarbons at low pressures where it is the metastable form of carbon.[1][2][3] However, growth rates were too slow to be of commercial interest. During the past decade energetically assisted growth methods have been developed which permit greatly enhanced growth rates.[4-7] Microwave plasma assisted chemical vapor deposition, hot filament assisted deposition and DC plasma assisted deposition are the most common methods employed. While details of the processes differ, some factors are common to all. For example, all processes operate in regimes where there are significant concentrations of atomic hydrogen and energetic carbon containing fragments. Growth conditions where there is high surface mobility are employed. This is usually achieved by using substrate temperatures in the range from 800 to 1000 C. Typical source gases are simple hydrocarbons, e.g., methane, diluted to 1% with hydrogen. Recent reviews describe the various processes and give access to the literature.[8][9][10]

Related research has led to the recognition of a new class of non—crystalline diamondlike carbons and hydrocarbons. They were originally called "diamondlike" because of their unusual hardness and chemical inertness.[11] These properties appear to arise from their unusually high proportion of sp^3 carbon sites compared to other amorphous carbons. The diamondlike phases are typically made by ion assisted processes, e.g., by direct deposition from low energy ion beams, by various sputtering processes and by RF self bias deposition. Again several reviews are available.[8][10][12][13]

2. CATEGORIZATION OF DIAMOND AND "DIAMONDLIKE" PHASES

The diamond and diamondlike phases are categorized by a method proposed by Angus.[10][14] This is shown in Figure 1, which is a plot of the number density, i.e., the number of atoms per cubic centimeter, versus the atomic fraction hydrogen. On this plot the various classes of hydrocarbons appear in tight groupings.

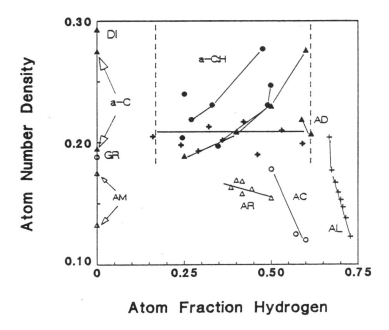

Atom Fraction Hydrogen

Figure 1.   Atom number density vs. atom fraction hydrogen.   The ordinate is expressed in gram—atoms/cm³, which is the total number of atoms per cubic centimeter divided by Avogadro's number.

Legend: DI, diamond; a—C, diamondlike carbon; a—C:H, diamondlike hydrocarbons; AC, oligomers of acetylene; AD, adamantanes; AL, n—alkanes, AM, amorphous trigonally bonded carbon; AR, polynuclear aromatics; GR, graphite. Filled symbols are used for the diamondlike hydrocarbons.[10]

The diamondlike hydrocarbons (a—C:H) fall in the range from approximately 0.20 to 0.60 atom fraction hydrogen and number densities greater than 0.19 gram—atoms/cm³.   The vertical dashed lines show the theoretical composition range predicted for fully constrained, random hydrocarbon networks.[18]

The diamondlike carbons (a—C) are a different class of material with little or no hydrogen and with atom number densities from approximately 0.19 to 0.28 gram—atom/cm³.

Two different types of diamondlike phases may be distinguished on Figure 1, viz., the amorphous diamondlike hydrocarbons (a—C:H) and the amorphous diamondlike carbons (a—C).*   The a—C:H fall in the region between crystalline diamond and the adamantanes at hydrogen atom fractions ranging from approximately 0.17 to 0.60 and atom number densities above 0.19.   (The adamantanes may be thought of as molecular diamonds containing 10 and 14 carbon atoms, saturated with hydrogen.)   The diamondlike carbons (a—C) contain very little hydrogen and have number densities between that of graphite (GR) and diamond (DI).

---

*   There is no uniformly accepted nomenclature for the diamondlike phases.   Every choice has some difficulties.   The use of the term "diamondlike," which was apparently coined by Aisenberg [11], is in very widespread use and is continued here as a generic descriptor for all carbon and hydrocarbon phases with atomic number densities greater than approximately 0.19 g atom/cm³.   It is appropriate, not only because of the hardness and density of these materials, but also because their properties appear to depend on the anomalous fraction of tetrahedrally coordinated (sp³) carbon sites.   The term a—C:H has been widely used for the diamondlike hydrocarbons and was chosen in analogy to the related material, amorphous hydrogenated silicon (a—Si:H).   No such agreement on terminology appears for the diamondlike carbons.   We arbitrarily chose the descriptor a—C despite the fact that it has sometimes been used to describe the amorphous carbons based on a trigonally bonded, sp², structure which are typically formed by evaporation.

Considerable understanding of the amorphous diamondlike hydrocarbons (a—C:H) has been achieved. Koidl and co—workers have been particularly active in characterization of these materials.[12 15] Robertson and Bredas have described their electronic and optical properties.[16 17] Angus and Jansen have shown that a—C:H acts as a random hydrocarbon network and have predicted the composition range over which the a—C:H phases are stable.[18] These predicted composition limits are shown as the vertical dashed lines on Figure 1. The a—C:H appear to be comprised of a random hydrocarbon network, within which are isolated clusters dominated by sp^2, trigonally coordinated, carbon atoms. These graphitic like clusters account for the observed optical absorption, but because they are isolated do not contribute to the conductivity. The atom number density of a—C:H increases with hydrogen content in contrast to the behavior of conventional hydrocarbons.

Much less is known about the structure and bonding within the amorphous diamondlike carbons (a—C). There is, however, much evidence that their extreme properties arise from an unusually high concentration of sp^3, tetrahedrally coordinated, carbon sites.[19-24]. (High number densities could not be achieved without significant concentration of the four—fold coordinated sp^3 carbon sites.) Further evidence of the lack of sp^2 carbon sites is obtained from EELS measurements which show the absence of π plasmon absorption.[20] A truly random carbon network comprised of only sp^2 and sp^3 carbon sites will be very highly strained and unlikely to be stable.[18] This implies that the structure in the diamondlike carbon phases involves some medium range ordering or micro—crystallinity.

The relationship between the diamondlike carbons (a—C) and true crystalline diamond is of great interest. One form of the a—C may be the extremely finely divided polycrystalline diamond grown by energetically assisted chemical vapor deposition at high supersaturations. This often takes the form of spherical, polycrystalline diamond balls. Other, unusual structures, perhaps including sp^1, doubly coordinated, carbon are also possible. Merz and Hoffman have proposed a number of unusual carbon networks that are still other possibilities.[25]

3. VAPOR DEPOSITED DIAMOND AS AN OPTICAL MATERIAL

3.1. Properties of diamond

Diamond has unusually high values of hardness, refractive index, strength, transparency and thermal conductivity. In addition, diamond has a very low thermal expansion coefficient and is chemically inert except in the presence of oxidizing agents at high temperatures. This unusual set of properties makes diamond a unique optical material. The physical properties of diamond have been compiled elsewhere [26] and its optical properties described in another paper in this symposium.[27] In Table 1 we give a brief summary of the most salient properties of natural type IIA diamond[26] and a comparison with the properties of vapor—grown diamond where possible.[8 10]

3.2. Optical applications of vapor—grown diamond: limitations and current status

The vapor growth of high quality optical grade diamond crystals by current techniques is limited by two factors: 1) the uncontrolled formation of independent diamond nuclei during growth and 2) the appearance of growth errors during extension of the lattice. The first leads to polycrystalline films with a typical crystal size of one to ten microns. The second leads directly to defect structures within the crystalline lattice. In order to grow the highest optical quality crystals from the vapor it will be necessary to control both the nucleation of new crystals and the growth errors on existing crystal faces.

Many electronic applications of vapor—grown diamond will require very high quality single crystals. On the other hand, many optical applications, for example as windows and mirror substrates, may tolerate polycrystalline films if scattering of photons and phonons by grain boundaries is not too severe. However, up to the present time the reported CVD diamond films are too rough and too absorbing for most applications. Typical polycrystalline diamond films grown by the hot filament or microwave process have surface roughnesses on the order of 0.5 microns. This will introduce enough scattering to preclude their application as optical

Table 1. Some Properties of Natural[26] and Vapor—Grown Diamond[8] [10]

	Type IIA	Vapor—grown
Hardness, GPa	≃90 *	80 — >90
Mass density, g/cm^3	3.515	2.8 — 3.5
Molar density, g atom/cm^3	0.293 *	0.23 — 0.29
Specific heat at 300 K, J/g	6.195	
Debye temperature, 273—1100 K	1860 ± 10 K	
Thermal conductivity at 298 K, w/cm K	≃20 *	10 — 20
Bulk modulus, N/m^2	4.4—5.9 x 10^{11}	
Compressibility, cm^2/kg	1.7 x 10^{-7} **	
Thermal exp. coeff. at 293 K, K^{-1}	0.8 x 10^{-6} ***	
Refractive index at 589.29 nm	2.41726	≃2.4
Dielectric constant at 300K	5.7 ± 0.05	≃5.7

*   higher than any other known material
**   lower than any other known material
***   lower than Invar

coatings except possibly at far infrared wavelengths. Physical polishing of polycrystalline diamond films to an optical smoothness is certainly a non—trivial matter and chemical etching techniques to achieve planarization may be more likely achieved.

Current techniques for the vapor deposition of diamond employ temperatures between 800 and 1000 C. These temperatures are too high for many substrates of interest and further limit the applications of vapor grown diamond. However, low energy ion beams, which can provide very high equivalent surface temperatures on cool substrates, have been used to produce microcrystalline diamond films.[28]

3.3. Long range applications of vapor grown diamond

Crystalline diamond grown by chemical vapor deposition (CVD) is of great interest because it may ultimately permit the growth of optical quality diamond in shapes and sizes that have not before been obtainable. Thin films, lenses, optical elements and large area, free standing instrument windows made of diamond may be feasible. Diamond's high thermal conductivity, strength and low thermal expansion coefficient give it exceptional resistance to thermal shock and it is therefore of potential interest as a substrate material for high power laser mirrors.

Natural, insulating diamond has been shown to be an excellent photoconductive switch for short wavelength laser sources.[29] [30] In the range from 0.22—0.6 micron, defect and impurity photo—absorption provides the switching mechanism. Moreover, since the absorption is weak, efficient volume excitation occurs. Sub—nanosecond, kilovolt switching has been reported for micro—joule optical pulses. Vapor deposited large single crystals might have significant advantages over natural stones because of reproducible levels of vacancy and nitrogen defect complexes as well as increased simplicity of doping.

4. DIAMOND GROWTH FROM THE VAPOR

4.1. Nucleation of new diamond crystals from the vapor

Very little is understood about the mechanism of nucleation of new diamond crystals from the vapor. Surface preparation, e.g., scratching the substrate with diamond powder or other abrasive, increases the nucleation rate dramatically, perhaps by transferring diamond fragments to the surface.[31] [32] The nucleation rate is also strongly dependent upon the type of substrate, e.g., nucleation rates are greater on boron than on other metallic or oxide substrates.[33] Finally, there is some indication that diamond crystals may actually nucleate in the vapor phase.[34][35] Matsumoto has proposed several cage hydrocarbon compounds as possible diamond precursors.[36] However, these compounds, e.g., adamantane, are not thermally stable at the growth temperatures. Carbon clusters stabilized by other elements, e.g., Si or B, may be more likely nuclei.

4.2. Vapor growth of diamond

Extension of an existing diamond lattice by addition of carbon atoms is a separate problem from the nucleation of new diamond crystals. Considerable effort has been expended attempting to identify a key pre—cursor species that adds onto the growing diamond surface. Harris [37] has recently shown that the two most likely species arriving at the surface from the plasma phase are the methyl group, CH_3, and acetylene, C_2H_2. The latter had been earlier proposed as the key intermediate by Frenklach and Spear.[38]

There is evidence that the growth mechanism does not involve the direct attachment of species from the vapor phase, but rather attachment of mobile adsorbed species on the surface.[33] These species diffuse on the surface until they find an energetically attractive surface site to which they become permanently bonded. The detailed atomic level structure of the diamond surface at the growth conditions must, therefore, play a key role in the growth process. For example, chemisorbed hydrogen is known to aid in maintaining the correct diamond lattice structure to the outermost surface layer.

Furthermore, simple arguments indicate that attachment of single carbon atoms will be energetically favored on {100} planes.[10] The critical nucleus size on a {100} plane is only a single carbon atom.[39] Rapid addition of single carbon atoms to {100} planes will lead to {111} growth surfaces. Subsequent growth of the diamond will require a more difficult formation of a two—dimensional, four—carbon—atom nucleus on the {111} surface. The most common growth error during extension of the diamond lattice appears to be the formation of two—dimensional "hexagonal" nuclei on {111} surfaces.[10] These nuclei are formed when the adatoms take the "boat" conformation rather than the "chair" conformation. This growth error is illustrated in Figures 2 and 3. Extension of a "hexagonal" nucleus across a {111} plane by layer growth will lead to a single twin plane. Two of these growth errors on adjacent (111) planes will lead to a stacking fault. Various types of line defects arise when cubic and hexagonal growth layers meet on a {111} plane. For example, if these layers meet along a <110> direction, a line defect comprised of trigonally bonded, sp^2 carbon sites can form.

The energy difference between the cubic and hexagonal two dimensional nuclei is quite small, on the order of 0.1 electron volt per carbon atom. Consequently, this growth error will easily occur, especially at high carbon supersaturations. It is significant that few stacking faults are found in natural diamond, which is believed to grow at very low supersaturations, somewhat more in high pressure synthetic diamond,[40] and the most in vapor—grown diamond grown at high supersaturations.

Another common problem with vapor grown diamond, especially diamond grown at high supersaturations, is the presence of a so—called "diamondlike" phase in addition to the crystalline diamond. This phase is identified by its characteristic broad Raman peak near 1500 cm^{-1}. The structural feature associated with this peak is not known with certainty. However, an attractive working hypothesis is simply that it arises from the network of dislocations and stacking faults that arise from the hexagonal growth errors occurring on {111} planes. Simple molecular models indicate that many types of structures can occur when the cubic and hexagonal layers meet. Some of these appear to be stabilized by three—fold coordinated (sp^2) carbon sites.

5. DIAMONDLIKE HYDROCARBONS (a—C:H)

5.1. Properties and growth of diamondlike hydrocarbons

The diamondlike hydrocarbons (a—C:H) are the hardest and most dense hydrocarbon known. They are extremely impermeable [18][41] and are often harder than silicon carbide. The hardness of a—C:H compared to diamond—cubic and zinc—blende solids is shown in Figure 4. The a—C:H may truly be thought of as a

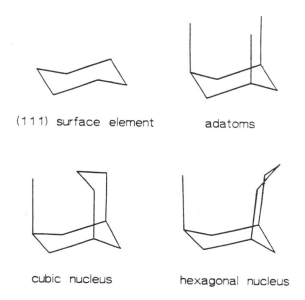

(111) surface element   adatoms

cubic nucleus   hexagonal nucleus

Figure 2.   Unreconstructed diamond surfaces and nucleation sites.

The figure shows an element of a {111} surface, an element of the surface with three carbon atoms (or methyl groups) attached, and the start of cubic and hexagonal nuclei on the surface element. Cubic and hexagonal nuclei differ only in next—nearest neighbor interactions. In a cubic nucleus the carbon—carbon bonds to the surface are in the staggered conformation so steric interactions with the surface plane are minimized. All six—membered rings are chairs. The hexagonal nucleus is formed when the carbon—carbon bonds are eclipsed, that is, the middle atom of the three—atom bridge is directly above an atom in the surface plane. This leads to a six—membered ring with the boat conformation.

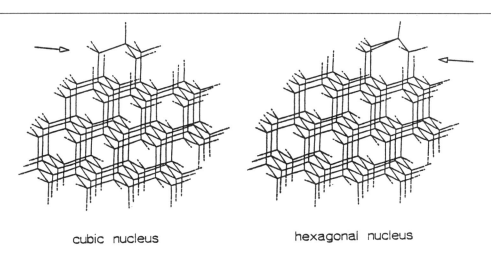

cubic nucleus   hexagonal nucleus

Figure 3.   Cubic and hexagonal nuclei on a {111} diamond surface. The 4—atom carbon atom nuclei on the larger diamond cluster are indicated by arrows. Extension of either nucleus across a {111} surface can be accomplished by addition of single—carbon and two—carbon atom species. Layer growth from a cubic nucleus in this manner will lead to an extension of the diamond—cubic lattice. Extension of a hexagonal nucleus across a {111} surface of a diamond—cubic lattice will lead to a twin plane.

Molecular mechanics energy minimization of the two structures in Figure 3 shows that the cubic nucleus is more stable than the hexagonal by approximately 0.1 electron volt per carbon atom.

"hydrocarbon ceramic." These unusual properties appear to arise because they film structure approaches the limiting case of a random network that is completely mechanically constrained.[18] In other words, the a—C:H are the maximally crosslinked random hydrocarbon polymer.

The a—C:H films can be easily grown on a wide variety of substrates by relatively simple deposition processes.[12] The films form when hydrocarbon ions with energies in the range from tens to several hundred electron volts fall on a substrate. A great advantage of a—C:H is that it is formed on cooled substrates. The films can be easily deposited on polymers for example.

The simplest deposition method is the RF self—bias method pioneered by Holland.[42] In this process an RF discharge is maintained through a capacitively coupled power supply. The large difference in mobilities betweens electrons and ions in the plasma leads to the spontaneous generation of a negative potential on the electrodes sufficient to maintain the zero current condition. Hydrocarbon ions from the plasma region are accelerated across the sheath and impinge upon the electrode forming the film. The a—C:H films can also be made using direct deposition from low energy beams of hydrocarbon ions.[12]

5.2. Optical applications of diamondlike hydrocarbons (a—C:H)

"Diamondlike" hydrocarbon films provide excellent protective, anti—reflection coatings in the infrared (6 — 12 microns). The combination of infrared transparency, good substrate adhesion, extreme abrasion resistance,[43] hardness, and a refractive index tunable around $\{n(Ge)\}^{0.5} = 2.0$ are nearly ideal for germanium infrared optical elements. Other substrates include solar cells[44] and aluminum mirrors used in thermal imaging.[45] Films of a—C:H can be used both as a single quarter—wave layer[46] and as the terminal layer in a broad—band multi—layer stack.[15] A quarter—wave coating on diamond may significantly reduce the reflectivity at $\simeq 1$ micron (Nd:YAG),[27] however the high absorption[47] ($\simeq 100$ cm^{-1}) in comparison with type IIA diamond ($\simeq 0.001$ cm^{-1}) and even type IIB ($\simeq 1$ cm^{-1}) clearly will limit the maximum optical fluence.

Because they are compressively stressed, the a—C:H films can increase the crack nucleation threshold[48][49] and perhaps the fracture toughness of brittle infrared window materials.[50] The a—C:H films are also under intensive study as possible tribological coatings for both optical and magnetic disks because of their hardness, smoothness and impermeability.

5.3. Applications of diamondlike hydrocarbons (a—C:H) in lithography

Diamondlike hydrocarbon films are now a material of choice for the fabrication of nanometer structures — particularly gratings and high aspect—ratio features.[51] The DLC films are used as bottom layer etch masks underneath an electron—beam resist in a bi—layer system. The diamondlike films have a high etch rate with O_2 reactive ion etching (preserving the fidelity of the pattern transfer from the top resist) and a low etch rate against the dry etch for the substrate (Si or GaAs), and the mechanical integrity to allow high aspect ratios. Patterns as small as 40 nm have been transferred to the substrate.[52]

Diamondlike films have also been used as a single layer self—developing resist upon exposure to 0.1—1.0 J/cm^2 pulses of the ArF excimer.[53] Good resolution of 130 nm lines has been achieved by single pulse, projection lithography.[54] In this application low hydrogen, lower stress films made from butane at high ($\simeq 1$ kv) biases appear to be advantageous.[55][56]

5.4. Limitations of diamondlike hydrocarbons (a—C:H)

The a—C:H films have some properties which limit their usefulness. The films are absorbing in the visible[12] with a typical absorption coefficient of 10^3 cm^{-1} at 500 nm. Unfortunately, the absorption coefficient and hardness both decrease as the hydrogen content of the films increases. The most transparent a—C:H films are found at hydrogen atom fractions approaching the theoretical limit of 0.615 where the films are the softest and the least abrasion resistant.

The a—C:H films are highly compressively stressed and the stress increases with hydrogen content (Figure 5). The stress levels are sufficiently high that thick films, i.e., greater than 1 micron, tend to delaminate from the substrate. The a—C:H films are also thermally unstable. Above about 400 C the films evolve hydrogen and revert to a more graphitic—type structure. The conductivity increases dramatically and the films become softer.

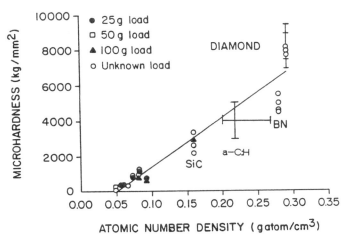

Figure 4.   Microhardness of diamondlike hydrocarbons (a—C:H) and solids with the diamond—cubic and zinc—blende structures versus atomic number density.

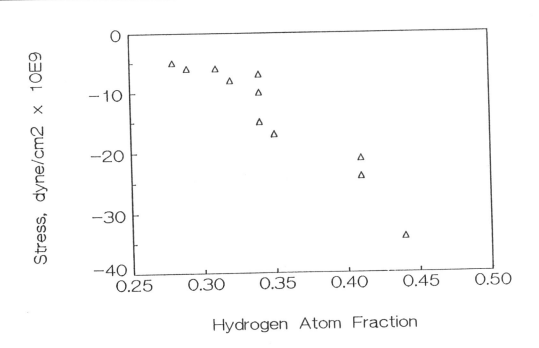

Figure 5.   Internal stress of diamondlike hydrocarbon films versus hydrogen atom fraction. Compressive stresses have a negative sign. The films were deposited on quartz wafers and the stresses measured by determining the deflection of the substrate. The hydrogen concentrations were measured using ^{15}N nuclear reaction analysis.

6. DIAMONDLIKE CARBONS (a—C)

Much less is known about the detailed structure of the amorphous diamondlike carbons (a—C) than about the diamondlike hydrocarbons (a—C:H). However, it is clear that the a—C contain little hydrogen and that a majority of the carbon sites are tetrahedrally coordinated sp^3 sites.[19-24] The a—C films can be made by variations of magnetron sputtering, by ion beam assisted processes and by condensation of carbon from energetic plasmas. The properties of these diamondlike carbons are so striking that they have been referred to as "amorphous diamondlike carbon (a—D)." [57]

The a—C films are reported to have exceptional hardness, but detailed quantitative data are sparse. The fact that the films can be made very smooth and can be formed on cold substrates makes them of great interest as protective coatings for temperature sensitive substrates. Furthermore, it appears that the a—C films are less compressively stressed than the a—C:H films and can therefore be deposited as thicker layers.

The films can be rather transparent or highly absorbing. The optical absorption may arise from sp^2 (graphitic) type clustering or on other, as yet unknown, defect structures. It is very likely that several types of complex carbon structures have all been classed under the general term "diamondlike carbons."

7. CONCLUSIONS AND SUMMARY

New optical technologies based on the chemical vapor deposition of crystalline diamond are being developed. The most interesting applications of vapor—grown diamond films await reductions in surface roughness and optical absorption. Massive research efforts in this country and abroad are being directed towards solving these problems.

The diamondlike hydrocarbons (a—C:H) should be recognized as a completely new class of maximally crosslinked hydrocarbon polymer. Because their properties can be varied by varying the amount of hydrogen and tertiary elements, these amorphous "hydrocarbon ceramics" show exceptional promise as protective coatings and diffusion barriers. They are already finding application as coatings on infrared optical elements and in photolithography.

The amorphous diamondlike carbons (a—C) are an extremely interesting, but poorly understood, class of new material with exceptional hardness and smoothness. They may be a form of micro—crystalline diamond or other, more complex, structures containing significant amounts of trigonally coordinated or even doubly coordinated carbon.

A summary of some possible applications of vapor—grown crystalline diamond films and of the diamondlike films is given in Figure 6.

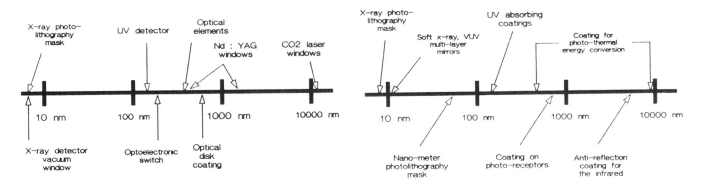

POTENTIAL APPLICATIONS
FOR VAPOR-GROWN DIAMOND

CURRENT AND POTENTIAL
APPLICATIONS FOR DIAMONDLIKE FILMS

Figure 6.   Potential applications of vapor-grown diamond; current and potential uses of diamondlike phases.

8. ACKNOWLEDGEMENTS

Mr. Hsiung Chen grew a—C:H films and Miss Susan Heidger made the compressive stress measurements. Mr. Fred Buck did the computer modelling of the diamond nuclei. Dr. Paul Schmidt made important contributions and suggestions throughout the work. William Lanford performed the hydrogen analyses.

9. REFERENCES

1.   W.G. Eversole, U.S. Patents 3, 030, 187 and 3, 030, 188 (1962).
2.   J.C. Angus, H.A. Will and W.S. Stanko, "Growth of Diamond Seed Crystals by Vapor Deposition," J. Appl. Phys. 39, 2915–2922 (1968).
3.   B.V. Deryagin, D.V. Fedoseev, V.M. Lukyanovich, D.V. Spitsyn, V.M.Ryabov and A.V. Lavrentev, "On needlelike crystal of diamond," Doklady Akad. Nauk SSSR 181, 1094–1096(1968); B.V. Deryagin et al., "Filamentary diamond crystals," J. Crystal Growth 2, 380–384 (1968).
4.   S. Matsumoto, Y. Sato, M. Kamo and N. Setaka, Jpn. "Vapor Deposition of Diamond Particles from Methane," J. Appl. Phys., Part 2, 21, L183–L185 (1982).
5.   S. Matsumoto, Y. Sato, M. Tsutsumi and N. Setaka, "Growth of diamond particles from methane—hydrogen gas," J. Mat. Sci. 17, 3106–3112 (1982).
6.   M. Kamo, Y. Sato, S. Matsumoto, and N. Setaka, "Diamond synthesis from gas phase in microwave plasma," J. Cryst. Growth 62, 642 (1983).
7.   Y. Matsui, S. Matsumoto and N. Setaka, "TEM—electron energy loss spectroscope study of the diamond particles prepared by the chemical vapor deposition from methane," J. Mat. Sci. Lett. 2, 532 (1983).
8.   R.C. DeVries, "Synthesis of diamond under metastable conditions," Ann. Rev. Mater. Sci. 17, 161 (1987).
9.   A.R. Badzian and R.C. DeVries, "Crystallization of diamond from the gas phase," Mat. Res. Bull. 23, 385 (1988).
10.   J.C. Angus and C.C. Hayman,"Metastable Growth of Diamond and 'Diamondlike' Phases," to be published, Science (August, 1988).
11.   S. Aisenberg and R. Chabot, "Ion—beam deposition of thin films of diamondlike carbon," J. Appl. Phys. 42, 2953–2958 (1971).

12.   J.C. Angus, P. Koidl and S. Domitz, "Carbon Thin Films." Chap. 4, in Plasma Deposition of Thin Films, J. Mort and F. Jansen, Eds., CRC Press, Boca Raton, Fl (1986).

13.   H. Tsai and D.B. Bogy, "Characterization of diamondlike carbon films and their application as overcoats on thin–film media for magnetic recording," J. Vac Sci. and Tech. A5, 3287–3312 (1987).

14.   J.C. Angus, "Empirical categorization and naming of 'diamond–like' carbon films," Thin Solid Films 142, 145–151 (1986).

15.   A. Bubenzer, B. Dischler, G. Brandt and P. Koidl, "rf–plasma deposited amorphous hydrogenated hard carbon thin films: preparation, properties, and applications," J. Appl. Phys. 54, 4590–4595 (1983).

16.   J. Robertson and E.P. O'Reilly, "Electronic and atomic structure of amorphous carbon," Phys. Rev. B35, 2946–2957 (1987); J. Robertson, "Amorphous carbon" Adv. Phys. 35, 317–374 (1986).

17.   J.L. Bredas and G.B. Street, "Electronic properties of amorphous carbon films," J. Phys. C. 18, L651 (1985).

18.   J.C. Angus and F. Jansen, "Dense 'diamondlike' hydrocarbons as random covalent networks," J. Vac. Sci. Tech. A6. 1778–1782 (May–June, 1988).

19.   V.E. Strel'nitskii, V.G. Padalka and S.I. Vakula, "Properties of the diamondlike carbon film produced by the condensation of a plasma stream with an rf potential," Sov. Phys. Tech. Phys. 23, 222–224 (1978).

20.   T. Miyazawa, S. Misawa, S. Yoshida and S. Gonda, "Preparation and structure of carbon films deposited by a mass separated C^+ ion beam," J. Appl. Phys. 55, 188–193 (1985).

21.   N. Savvides, "Optical constants and associated functions of metastable diamondlike amorphous carbon films in the energy range 0.5–7.3 eV," J. Appl. Phys. 59, 4133–4145 (1986).

22.   N. Savvides and B. Window, "Diamondlike amorphous carbon films prepared by magnetron sputtering of graphite," J. Vac. Sci. Tech. A3, 2386–2390 (1985).

23.   S.M. Rossnagel, M.A. Russak and J.J. Cuomo, "Pressure and plasma effects on the properties of magnetron sputtered carbon films," J. Vac. Sci. Tech. A5, 2150 (1987).

24.   A. Grill, B.S. Meyerson, V.V. Patel, J.A. Reimer and M.A. Petrick, "Inhomogeneous carbon bonding in hydrogenated amorphous carbon films," J. Appl. Phys. 61, 2874–2877 (1987).

25.   K.M. Merz, R. Hoffman and A.T. Balaban, "3,4–connected carbon nets: through–space and through–bond interactions in the solid state," J. Am. Chem. Soc. 109, 6742–6751 (1987).

26.   J.E. Field Properties of Diamond, Academic Press, London (1979).

27.   M. Seal and W.J.P. van Enckevort, "Application of diamond in optics," this symposium.

28.   M. Kitabatake and K. Wasa, "Growth of diamond at room temperature by an ion–beam sputter deposition under hydrogen–ion bombardment," J. Appl. Phys. 58, 1693–1695 (1985).

29.   P.S. Panchhi and H.M. Van Driel, "Picosecond optoelectronic switching in insulating diamond," IEEE Journal of Quantum Electronics, QE–22, No. 1, 101–107 (January, 1986)

30.   J. Glinski, X.J. Gu, R.F. Code, H.M. Van Driel, "Space–charge–induced optoelectronic switching in IIa diamond," Appl. Phys. Lett. 45, 3, 260–262 (1 August 1984).

31.   K. Mitsuda, Y. Kojima, T. Yoshida and K. Akashi, "The growth of diamond in microwave plasma at low pressure," J. Mat. Sci. 22, 1557–1562 (1987).

32.   C.P. Chang, D.L. Flamm, D.E. Ibbotson and J.A. Mucha, "Diamond crystal growth by plasma chemical vapor deposition," J. Appl. Phys. 63, 1744–1748 (1988).

33.   T. Anthony, "The growth rate of filament–assisted plasma CVD diamond," Symposium N, Fall Meeting, Materials Research Soc. Meeting, Boston, Nov. 30–Dec. 5, 1987.

34.   S. Mitura, "Nucleation of diamond powder particles in an RF methane plasma," J. Cryst. Growth 80, 417–423 (1987).

35.   B. Deryagin, "Development of the science of diamond deposition from the vapor," Symposium N, Materials Research Soc. Meeting, Boston, Nov. 30–Dec. 5, 1987.

36.   S. Matsumoto and Y. Matsui, "Electron microscopic observation of diamond particles grown from the vapour phase," J. Mat. Sci. 18, 1785–1793 (1983).

37.   S. Harris, ONR/SDIO Diamond Technology Symposium, Arlington, VA, July, 1988; to be published Appl. Phys. Lett.

38.   M. Frenklach and K.E. Spear, "Growth mechanism of vapor deposited diamond," J. Mater. Res. 3, 133–140 (1988).

39. J.W. Faust, Jr. and H.G. John, "Growth facets on III–V intermetallic compoounds,"J. Phys. Chem. Solids 23, 1119–1122 (1962).

40 A.R. Lang, "Internal Structure," in Properties of Diamond, Academic Press, London, 425–469 (1979).

41. Ch. Wild and P. Koidl, "Thermal gas effusion from hydrogenated amorphous carbon films," Appl. Phys. Lett. 51, 1506–1508 (1987).

42. L. Holland and S.M. Ojha, "Deposition of hard and insulating carbonaceous films on an r.f. target in a butane plasma," Thin Solid Films 38, L17–L19 (1976); S.M. Ojha and L. Holland, Some characteristics of hard carbonaceous films," Thin Solid Films 40, L31–L32 (1977).

43. R.J. King, D.E. Putland and S.P. Talim, "The abrasion of optical coatings and its assessment," J. of Physics E: Scientific Instrumentation, 21, 40–46 (1988)

44. H. Vora and T.J. Moravec, "Structural investigation of thin films of diamondlike carbon," J. Appl. Phys. 52, 6151–6157 (1981).

45. A.H. Lettington, "Optical and other applications of hard carbon as a durable thin film material," Proc. European Materials Research Society Meeting, Vol XVII, 359–369 (June, 1987).

46. K. Enke, "Hard carbon layers for wear protection and antireflection purposes of infrared devices," Appl. Optics 24, 508–512 (1985).

47. B. Dischler, A. Bubenzer, P. Koidl, "Bonding in hydrogenated hard carbon studied by optical spectroscopy," Solid State Commun. 48, 105–107 (1983); J. Wagner, Ch. Wild, F. Pohl and P. Koidl, "Optical studies of hydrogenated amorphous carbon plasma deposition," Appl. Phys. Lett. 48, 106–108 (1986);

48. S. Van Der Zwaag, J.E. Field, "The Effect of Thin Hard Coatings on the Hertzian Stress Field," Phil. Mag. A 46, 133–150 (1982) and references therein.

49. G. Gille, "Strength of thin films and coatings," Current Topics in Matls. Sci. 12, 420–449 (1985).

50. P.H. Kobrin and A.B. Harker, "Compressive thin films for increased fracture toughness," SPIE 683, Infrared and Optical Transmitting Materials 139, (1986)

51. J.H. Wernick, "Some Electronic and Optical Applications of Diamond–like Films", personal communication.

52. M. Kakuchi, M. Hikita and T. Tamamura, "Amorphous carbon films as resist masks with high ion etching resistance for nanometer lithography," Appl. Phys. Lett. 48, 835–837 (1986)

53. M. Rothschild, C. Arnone and D.J. Ehrlich, "Excimer–laser etching of diamond and hard carbon films by direct writing and optical projection," J. Vac. Sci. and Tech. B4, 310–314 (1986).

54. M. Rothschild, D.J. Ehrlich, "Attainment of 0.13 micron lines and spaces by excimer projection lighography in 'diamond–like' carbon resist," J. Vac. Sci. and Tech. B 5, 389–390 (1987).

55. M. Rothschild and D.J. Ehrlich, "Critical review: a review of excimer laser projection lithography," J. Vac. Sci. and Tech. B 6, 1–17 (1988).

56. A.R. Nyaeish, R.E. Kirby, F.K. King and E.L. Garwin, J. Vac. Sci. and Tech. A 3,610–613 (1985).

57. S.D. Berger, D.R. McKenzie and P.J. Martin, "EELS analysis of vacuum arc–deposited diamond–like films," Phil. Mag. Lett. 57, 285–290 (1988).

Invited Paper

Crystallization of Diamonds by Microwave Plasma Assisted Chemical Vapor Deposition

Andrzej R. Badzian, Teresa Badzian and Dave Pickrell

Materials Research Laboratory
The Pennsylvania State University
University Park, PA 16802

ABSTRACT

Chemically vapor deposited diamond films have a potential for optical applications but so far only translucent films have been produced. The reason for the low transparency is considered to be related to the polycrystalline nature of the films and to defects. To elucidate the problem and indicate ways of improving film quality, the growth mechanism of CVD diamond must be understood. In microwave plasma assisted CVD the plasma is induced by microwaves in a mixture of hydrogen and methane. A high concentration of atomic hydrogen, much above the thermal equilibrium concentration, helps to prevent codeposition of graphite. The growth of diamond occurs in a narrow temperature range (900-1000°C) as a result of the interplay of several phenomena.

1. INTRODUCTION

Our interest in the microwave plasma assisted chemical vapor deposition (MPACVD) of diamond was stimulated by a paper of M. Kamo, et al. (1). The preliminary results of our work were reported at the 1986 SPIE meeting (2), and a general model for the nucleation and growth of diamond was developed to aid in determining optimal conditions for diamond crystallization (3,4). The influence of process parameters on the perfection of diamond crystals and films has been subsequently reported (5,6,7).

In this paper we will discuss the phenomena involved in the diamond growth process and the factors which limit the transparency of diamond films. We will then describe attempts to minimize the defects formed in films by depositing onto substrates downstream of the luminous plasma. Finally, the salient features of typical Raman spectra for diamond films will be discussed in terms of the information afforded about the structure of these films.

2. CVD DIAMOND CRYSTALLIZATION

Diamond crystallizes from the gas phase in a narrow temperature range. In chemical vapor deposition of diamond, crystals with the most perfect habit are formed from 950-1000°C. In this range the greatest lattice perfection is also achieved as revealed by structural characterization of deposited films. From plots of growth rate versus temperature, for various supersaturations of hydrocarbon species, it is observed that a maximum in diamond growth rate is obtained between 950 and 1000°C as well.

Our previously proposed model on the mechanism of diamond growth, which was found to be very useful in designing deposition experiments, stresses the complexity of the diamond growth process and describes it as an interplay of several phenomena. Chemical reactions between hydrocarbon species and atomic hydrogen occur in the boundary layer of gas adjacent to the substrate surface. The density of gas in this region is greater than that of the bulk gas because of the potential exerted upon it by the surface. The growing diamond surface varies from that of bulk diamond because of several phenomena occurring in the 900-1000°C temperature range. First, the chemisorbed hydrogen begins to desorb at approximately 900°C, and the desorption/adsorption reaction reaches an equilibrium surface coverage for a given temperature. Therefore the C-H bond weakens with increasing temperature in this range. Secondly, around a temperature of 950°C the precursors of surface reconstruction are expected to cause the carbon atoms on the diamond surface to vibrate with a greater amplitude. Surface reconstruction of clean diamond surfaces in high vacuum has been found to occur around 950°C, and although the CVD growth conditions do not involve high vacuum, this phenomena is felt to enhance the mobility of carbon atoms on the diamond surface.

Another phenomena which may be involved in the diamond growth process is the surface fluctuation of solids described recently by P.M. Horn (8). He found that Ag and Au (110) surfaces undergo a reversible equilibrium phase transition from smooth to rough with increasing temperature. The dynamic surface roughness was probed by measuring diffuse x-ray intensity in the vicinity of the (110) forbidden reflection. It was suggested that this surface transition can be related to crystal growth from the vapor phase. The roughening temperature for Ag was found to be about 500°C (melting point 960°C), with regular crystal growing above this temperature and dendrites forming below. We suggest that a similar situation may occur in diamond growth from the vapor phase, and that this may also enhance the mobility of carbon atoms and other species on the surfaces of growing diamond crystals.

Thus far in the CVD of diamonds, regular cuboctahedral crystals have been grown. Some crystals and films with high structural perfection and low hydrogen content (400 ppm) have been obtained. Considerable improvements in the growth rate of diamond have also been achieved. However, obstacles to the optical application of diamond films still exist. CVD diamond films are translucent because of the polycrystalline structure of films, amorphous carbon at the grain boundaries, and the defects within crystals. The occurrence of amorphous carbon and defects is due to the formation of graphite nuclei during diamond growth. It is therefore important to develop an understanding of the phenomena occuring at the surface during the growth process and how they influence the competing nucleation process between diamond and graphite.

3. DEPOSITION UNDER THE PLASMA

To improve the quality of diamond films in order to meet the requirements for optical applications, the MPACVD diamond method was reevaluated. Up to now, the process parameters most studied are the temperature, CH_4 concentration in H_2, total pressure and flow rate. Experiments are run in the geometrical center of the microwave cavity. We have begun to study the deposition of diamond films on substrates downstream of the plasma ball (position B, Fig. 1) to determine if the formation of amorphous carbon at the grain boundaries can be eliminated and the defect density within crystals can be reduced. It has been determined that the etching rate of a graphite rod immersed in the plasma varies along its length. The greatest etching occurs at the center of the cavity and just below the plasma ball, below the lower plane of the waveguide (see Figure 1). Active etching of graphite is desirable to suppress the nucleation of this phase. Therefore the region just below the plasma may allow diamond growth external to the harsh environment of the plasma minimizing the defects formed.

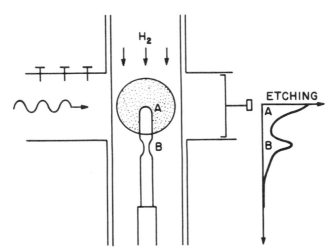

Fig. 1   Etching of graphite by hydrogen in microwave tube appliator. Etching rate is indicated by the adjacent graph. The two regions of intense etching are indicated by points A and B.

Independently heating the substrate, with a resistance heater from below, will allow the decoupling of substrate parameters from plasma parameters and facilitate understanding of the growth process. Issues to be addressed in this work include the temperature dependence of diamond growth, the lifetimes of active species involved in diamond growth, and the formation of amorphous carbon films at temperatures below that which is optimal for diamond growth. Preliminary experiments using a platinum heater and fused SiO_2 substrates have resulted in the deposition of transulcent films below the plasma. These films are more transparent and uniform than corresponding films deposited on fused SiO_2 immersed in the plasma. The Raman spectra varies across the film surfaces, but diamond crystals have been observed and identified. Work along these lines continues.

A chemical transport reaction method has been shown to be very effective in producing highly perfect single crystal diamond films on diamond, growth on diamond seeds and polycrystalline films on various substrates (9). In this technique, the carbon source (graphite) and substrate were held in a closed tube filled with hydrogen, and a temperature gradient was maintained between the source and substrate. We tested a similar configuration in a MPACVD tube reactor (Fig. 2a). A graphite foil sample holder was placed in the tube such that the top of the holder was in the center of the plasma, the substrate was held slightly below the plasma, and hydrogen flowed past the top of the holder to the substrate. The top of the holder, which was held at a temperature of 1300°C, was etched by the plasma and the hydrocarbon species formed were subsequently transported down to the substrate. The substrate was heated by conduction through the graphite to a temperature of 600-800°C. The amorphous films deposited on fused SiO_2 substrates were transparent, hard and gave a Raman spectra with no peaks in the range of 1200-1600 cm^{-1}. The absorption spectrum of a film for the visible and UV range is shown in Figure 3.

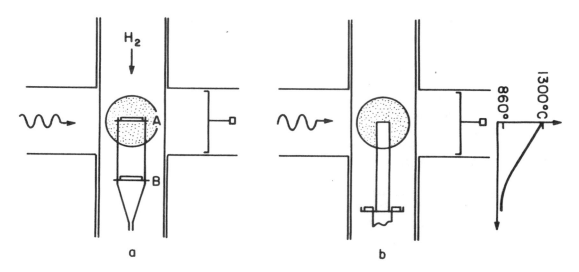

Fig. 2   Graphite susceptors designed to study the chemical transport reaction method in a microwave plasma system. Etched graphite from the top of the susceptors forms hydrocarbon species which are transported to the substrate. The temperature gradient along the graphite strip is shown.

A slightly different experimental arrangement is shown in Figure 2b. Again a graphite strip was immersed in the center of the plasma and the three substrates, Si, Mo and Fe, were held on a plate below the plasma. During deposition the top of the graphite strip was held at 1300°C while the Si substrate temperature was measured at 860°C. On the region of the graphite strip slightly below the plasma, a hard transparent deposit was obtained. Diamond microcrystals nucleated on the silicon substrate, a thick black soot was formed on the Fe, and no deposit was observed on the Mo. These results illustrate the differing behaviors of various substrates in the microwave system, and indicate the importance of growth mechanisms rather than thermodynamic factors in this situation.

4. CVD DIAMOND DEFECTS

As mentioned above, polycrystalline diamond films produced by MPACVD are translucent and greyish in color. Scattering from the grain boundaries is a major contributor to the lack of transparency. Amorphous carbon at the grain boundaries enhances light scattering, and can be detected by electrical resistivity measurements. In some highly oriented films this value can drop to $10^3 \Omega cm$. Planar defects within diamond crystals, seen externally on crystals as twins and overgrowths, also lower the transparency of films. These defects can be observed in TEM images of the {111} diamond planes. The defect density increases with supersaturation of carbon species (10).

An exact description of the planar defects on the atomic scale is not available at present. A crystallographic approach suggests that these defects are stacking faults on the {111} planes where there is a disturbance of the ABCABC sequence.

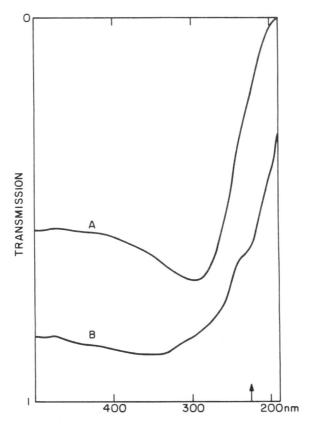

Fig. 3   Optical absorption spectra of: (a) Amorphous carbon film on fused silica deposited on stage B of susceptor shown in Fig. 2a, at a temperature from 600-700°C and a hydrogen pressure of 40 torr, (b) fused silica substrate.

Nucleation theory suggests the possibility of graphite nuclei formation on the {111} planes (11). Such flat nuclei can be covered by diamond planes and form various sizes of planar defects. These can be very flat in the <0001> direction and extended in the <1010> direction or vice versa. A great variety of defective diamond forms have been observed during our research with MPACVD. Raman spectroscopy has been shown to be an effective tool for studying these defects, however, this method is not sufficient to fully characterize them. Raman spectroscopy is sensitive to graphitic defects in diamond films with their signal being much enhanced over that from the ordered diamond regions. Therefore it is not as useful a technique for analyzing small regions of ordered diamond as electron diffraction is. We will attempt in the following to describe the various features of typical Raman spectra and relate that to the film structure.

Figure 4 shows a Raman spectrum for a material which contains separate domains of sp^3 and sp^2 bonding networks. The nature of this phase is not well understood. With single crystal graphite, only one coupling is allowed which results in the peak at 1580cm^{-1}. The peak at 1355cm^{-1} is related to the size of (0001) planes of graphite (12). Small crystallites produce many dangling bonds at the (0001) plane boundaries. Figure 5 shows that the Raman spectra of "white soot" and "black soot" are very similar. White soot behaves like a white powder, and scatters light strongly so that the background in Fig. 5b is high and no diamond peak appears. However, the electron diffraction pattern of white soot clearly shows the presence of the diamond phase and diffuse scattering in the vicinity of the primary beam. Black soot on the other hand is shown to be a highly disordered graphitic material by electron diffraction. Diamond networks in the white soot are so small and disturbed that they are not detected by Raman spectroscopy, which has a very small cross section for diamond in this case. So white soot represents a combination of sp^3 and sp^2 bonding networks. The sp^2 phase exists as separate domains or grains in which the ratio of dimensions <0001>/<1010> is small. The sp^2 type of bonding can also exist inside diamond networks. When domains of sp^3 and sp^2 become small (approximately 100Å), the Raman spectrum is like that shown in Fig. 5b. This phase can become white but in such a situation the sp^2 bonding would be different from that encountered in graphite.

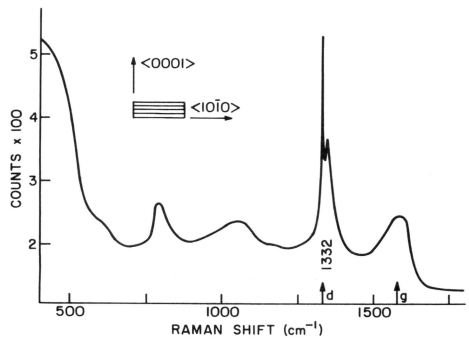

Fig. 4   Raman spectrum of a film deposited onto fused silica under the plasma. The substrate rested on a platinum wire resistance heater. The heater temperature was maintained at approximately 1100°C during the deposition. The CH_4 concentration was 0.3% and the total pressure was 90 torr.

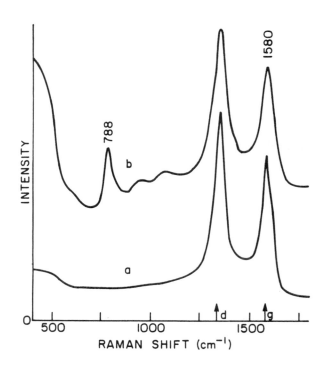

Fig. 5   Raman spectra of: (a) Black soot deposited onto Si while on stage A of susceptor in Fig. 2a with the following process conditions: $2\%CH_4$, $20\%Ar$, $78\%H_2$, 950°C substrate temperature and 40 torr total pressure. (b) White soot deposited onto fused SiO_2 also on stage A with the following conditions; $1\%CH_4$, $99\%H_2$, 935°C substrate temperature and 80 torr total pressure.

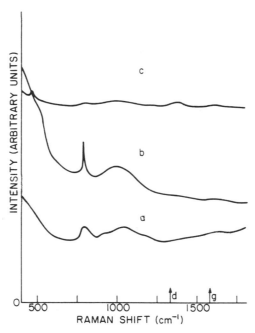

Fig. 6   Raman spectra of: (a) fused silica substrate, (b) amorphous carbon film on silica whose adsorption spectra is shown in Fig. 3 (c) amorphous carbon film on silica deposited in an H_2 atmosphere at 40 torr with a substrate temperature of 600-700°C.

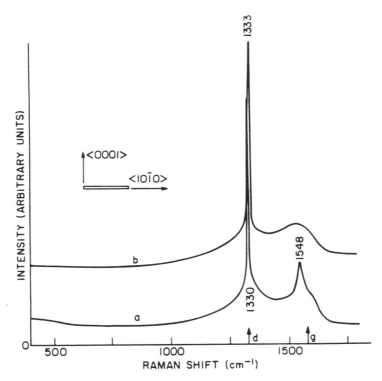

Fig. 7   Raman spectra of isolated diamond crystals: a) deposited onto fused silica below the plasma resting on a platinum heater: heater temperature 1100°C (the substrate temperature was not measured), 0.5% CH_4, 90torr total pressure; b) deposited onto a silicon substrate on stage A of susceptor in Fig. 2a with the following process conditions: 1% CH_4, 99% H_2, 80 torr total pressure and a substrate temperature of 985°C.

We believe that the diamond and graphite structures can interpenetrate such that graphitic planes are inserted between the {111} planes of diamond. We also think that the interplanar distance between graphitic planes and diamond {111} planes can be smaller than the interplanar distance in graphite (3.3Å). This graphitic-diamond plane distance is between 3.3Å and 2.06Å, the latter being the spacing between {111} planes in diamond. But decreasing the interplanar distances will cause atomic displacements, changing the sp^2 hybridization from that proper to the graphite structure and probably rearranging the spatial distributions of π electrons. As a consequence of this the optical properties are changed from that which is found in graphite to that which is observed in these films.

The intensity ratio of the 1355 to 1580 peaks and their profiles depend on process parameters such as temperature and CH_4 concentration. The position of the 1355 peak of white soot is usually shifted in the range of 1345-1355cm^{-1}. As the deposition temperature is lowered, the 1355 and 1580 peaks drop and eventually disappear (see Figure 6). Films deposited at low temperature are amorphous and transparent. We think white soot (Fig. 5b) is a intermediate stage of mixing of sp^3 and sp^2 bonding networks whereas amorphous films (Fig. 6b,c) contain sp^3 and sp^2 mixed on the nanometer scale.

When conditions are close to that which is optimal for diamond crystallization (5), the shape of the "graphitic" humps in the range of 1340-1600cm^{-1} change to produce a maximum between 1355 and 1580cm^{-1}. The spectra shown in Fig. 7 were obtained from isolated crystals with the help of a microfocus Raman spectrometer. Such spectra were also recorded from isolated diamond crystals grown on a graphite substrate. The most interesting feature to note is the peak in Fig. 7a which is shifted from the graphite position. As an explanation we suggest that this peak corresponds to a planar graphitic defect in the diamond matrix, parallel to the {111} planes, which is very thin in the <0001> direction (or in <111> of diamond) and extended in the <1010> direction. We believe that the sharpness of this 1548cm peak is due to a narrow size distribution of planar defects in these crystals. The sharp 1548 cm^{-4} peak represents a particular frequently occurring size of planar defect superimposed on the broader distribution of planar defects. On the other hand, the broad maximum in Fig. 7b corresponds to a wide variation of defect sizes in those crystals. As the CH_4 concentration in the plasma is reduced the thicknesses of these planar defects drops. At higher supersaturation of carbon species the graphitic layer becomes thicker in the <0001> direction. Defects in CVD diamond crystals are not independent, but can interact during the growth process. Our observations indicate that impurities like Si or N increase the density of planar defects.

5. DISCUSSION

Diamond is an excellent window material, however only natural diamond crystals belonging to groups IIa and IIb exhibit high optical transparency, and such crystals are rare in nature. Of all natural diamonds, 98% belong to group Ia which are not transparent in the U.V. region. Impurities and inclusions in natural diamond crystals produce optically active centers which limit their transparency. More than 50 such centers are currently known (13).

The optical quality of CVD diamond films is inferior to that of single crystal natural diamonds and is more comparable to that of ballas bodies which are naturally occurring polycrystalline aggregates of diamond. In our first paper on CVD diamond we discussed crystal defects as an obstacle to the preparation of high quality optical films (2). Today it appears that impurities in diamond films to some extent can be avoided but that conucleation of graphite is still a problem.

A series of experiments are being considered to determine the amount to which graphite defects can be reduced in MPACVD films.

* Epitaxial growth of diamond on diamond substrates

* Polycrystalline diamond films grown under the plasma

* Amorphous carbon films prepared at a temperature from 600°C-900°C which have mixed sp^3/sp^2 bonding

6. CONCLUSIONS

1. Diamond films and crystals prepared by MPACVD are relative structurally perfect and chemically pure.

2. CVD diamond films are translucent because of strong scattering at the grain boundaries which is enhanced by amorphous carbon present in these regions.

3. CVD diamond crystals contain planar defects as a result of conucleation of graphite during growth. Such defects have a wide range of dimensions and properties. It was proposed that broad peaks in the Raman spectra at 1355 and 1580cm-1 are related to the graphite inclusions (in the diamond matrix and/or between diamond grains) which are thick in the<0001> direction and short in the <1010> direction. The hump between the 1355 and 1580 cm-1 is related to graphitic defects which are thin in the <0001> direction and extended in the <1010> direction.

7. ACKNOWLEDGEMENTS

This work was supported in part by the Office of Naval Research and the Diamond and Related Materials Consortium. We extend our thanks to Professors Rustum Roy and Russell Messier. We also express our gratitude to Diane Knight for her assistance with the Raman Spectroscopy measurements, and Professor Larry Pillione for his helpful discussions.

8. REFERENCES

1. M. Kamo, Y. Sato, S. Matsumoto and N. Setaka, "Diamond Synthesis from Gas Phase in Microwave Plasma," J. Cryst. Growth, Vol. 62, pp. 642-644, 1983.
2. A.R. Badzian, B. Simonton, T. Badzian, R. Messier, K.E. Spear and R. Roy, "Vapor Deposition Synthesis of Diamond Films," Proceedings SPIE, 683, (Infrared and Optical Transmitting Materials), pp. 127-138, 1986.
3. A.R. Badzian, "Defect Structure of Synthetic Diamond and Related Phases," Advances in X-Ray Analysis Volume 31, pp. 113-128, Eds. C.S. Barrett et al., Proceedings of Annual Conference on Applications of X-Ray Analysis, 1987.
4. A.R. Badzian and R. DeVries, "Crystallization of Diamond from the Gas Phase, (Part 1)," Mat. Res. Bull., 23(3), 385-400 (1988).
5. A.R. Badzian, T. Badzian, R. Roy, R. Messier and K.E. Spear, "Crystallization of Diamond Crystals and Films by Microwave Assisted CVD (Part II)," Mat. Res. Bull., 23(4), 531-548, 1988.
6. A.R. Badzian and T. Badzian, "Surface Phenomena Related to Growth Mechanisms in CVD Diamonds," Diamond and Diamond-Like Materials Synthesis, G.H. Johnson, A.R. Badzian and M.W. Geis, eds. (Materials Research Society, 1988), EA-15 pp. 27-32.
7. A.R. Badzian and T. Badzian, "Nucleation and Growth Phenomena in CVD Diamond Coatings," Thin Solid Films (in press).
8. P.M. Horn, "Do Solids have Surfaces," Annual Meeting American Crystallographic Association, Philadelphia June 26-July 1, 1988. Paper WA2.
9. B.V. Spitsyn, "On Thermodynamics and Kinetics of Chemical Vapor Deposition of Diamond," Jn:IV USSR Conf. on Crystal Growth (in Russian) Yereva University Publishing House Yerevan, Part 1, 97-100, 1972.
10. Wei Zhu, A. Badzian and R. Messier, "TEM Studies of CVD Diamond Films," in preparation.
11. B.V. Derjaguin and D.V. Fedoseev, "Growth of Diamond and Graphite from Gas Phase," (in Russian) Nauka, Moscow, pp. 114, 1977.
12. F. Tuinstra and J.L. Koening, "Raman Spectrum of Graphite," J. Chem. Phys. Vol. 53(3), 1126-1130, 1970.
13. N.W. Novikov, Physical Properties of Diamond, Naukova Dumka, Kiev 1987 (in Russian).

SESSION 2

Deposition and Processing II

Chair
Andrzej R. Badzian
Pennsylvania State University

Growth of diamond films by hot filament chemical vapor deposition

Edward N. Farabaugh, Albert Feldman and Lawrence H. Robins

Ceramics Division
Institute for Materials Science and Engineering

and

Edgar S. Etz

Gas and Particulate Science Division
Center for Analytical Chemistry

National Bureau of Standards, Gaithersburg, Maryland 20899

ABSTRACT

Results of research on diamond films grown by the hot filament chemical vapor deposition process are discussed. The parameters for film deposition have been surveyed and the conditions for routine and reproducible film formation established for our deposition system. These were: 800°C substrate temperature, 52-78 sccm flow rate, 5×10^3 Pa deposition pressure and 99.5% H_2, 0.5% CH_4 gas composition. Characterization of the deposited films has been accomplished with scanning electron microscopy (SEM), x-ray diffraction (XRD), Auger electron spectroscopy (AES), electron energy loss spectroscopy (EELS) and Raman spectroscopy; and the presence of the diamond phase was verified. Initial depositions on Si and Al_2O_3 substrates resulted in individual diamond particles showing the distinct diamond morphology. These particles, when examined by Raman microprobe spectroscopy, displayed the diamond spectra. Subsequently, continuous diamond films were formed after pretreating the substrates by rubbing with 1 μm diamond abrasive before deposition. Films, all shown to be diamond, grown on fused silica, polycrystalline SiC and different orientations of single crystal Si (all pretreated), exhibit very similar surface topography and x-ray diffraction patterns. Additionally, x-ray diffraction shows no preferred orientation in the films. Mechanical surface measurements on a 5 μm thick film grown on Si show that the films possess an average surface roughness of 0.4 μm and a peak-to-valley roughness of 2.5 μm. Films deposited on optically clear fused silica substrates could be seen to be water white, suggesting no significant optical absorption. Considerable optical scatter, however, is present due to the roughness of the film surface. Deposition rates were of the order of 0.1 μm/hr, with the fastest apparent growth occurring on pretreated polycrystalline SiC substrates.

1. INTRODUCTION

Diamond films have attracted the attention of investigators because of diamond's unique physical, mechanical and chemical properties. As well as being chemically inert, diamond is the hardest known material and possesses the highest electrical conductivity, high thermal conductivity and optical transparency.[1] These properties offer the possible application of diamond films in the areas of protective optical coatings, wear resistant coatings, high temperature electronic devices and heat conductors.

Early publication of reports of successful diamond film deposition were made by Derjaguin and Fedoseev.[2], Spitsyn et al[3], and Matsumoto et al[4]. Several deposition methods, based on chemical vapor deposition (CVD), can now be successfully employed to produce diamond particles and continuous films. These methods include hot filament CVD,[4] election assisted CVD,[5,6] DC discharge CVD[7], RF discharge CVD,[8] microwave plasma-assisted CVD[9], laser-assisted deposition,[10] and remote plasma CVD.[11] In this paper results of research carried out at the National Bureau of Standards on the growth of diamond films using the hot filament CVD technique will be presented. Characterization of the films has been accomplished using x-ray diffraction (XRD), scanning electron microscopy (SEM), Auger electron spectroscopy (AES), electron energy loss spectroscopy (EELS), and Raman spectroscopy.

2. GROWTH AND CHARACTERIZATION OF DIAMOND FILMS

2.1 Initial depositions

The hot filament CVD apparatus, modified from that previously described[12], is shown in Figure 1. It consists of a fused silica tube which serves as the reaction chamber. The tube encloses a substrate holder, filament holder and filament, and tubing which directs a hydrogen-methane gas mixture over the heated filament onto the substrate. The tube is sealed to a base plate and the system is pumped by a mechanical pump. The fused quartz tube is inserted into a tube furnace which heats the entire reaction chamber to the desired deposition temperature. The temperature of the substrate is monitored via a thermocouple mounted in the substrate holder and is in turn used to control the furnace temperature by means of a feedback loop. External to the reaction chamber are mass flow controllers which are used to maintain specific gas mixtures and flow, both thermocouple and diaphragm vacuum gauges to read the pressure in the reaction chamber and a throttling valve in the pumping line to control the deposition pressure. Deposition parameters which were determined to yield diamond films were: substrate temperatures, 600°-850°C; deposition pressure 5 x 10^3 Pa; gas composition 99.5% H$_2$, 0.5% CH$_4$; and flow rate, 52-78 sccm. The filament was a 5 turn helix made of 0.125 mm (0.005 in) tungsten wire which was cured for several hours under operating conditions without the substrate being present.

The first depositions on Al$_2$O$_3$ and Si single crystal substrates resulted in only growth of individual diamond particles. SEM micrographs of typical particles are shown in Figure 2. Shown are (a) a twinned icosahedron, (b) a combination of {100} and {111} forms with {100} slightly dominant, and (c) nearly equal combination of {100} and {111} forms. These particles were usually well defined and of the order of 10 μm in their largest dimension. Because of their small size and lack of continuous film formation, no confirming XRD patterns could be taken. However, micro-Raman spectra were taken of the largest of the particles and a typical spectrum is shown in Figure 3. The spectrum consists of a sharp Raman line with very little background centered at 1322 cm^{-1}, which is 10 cm^{-1} below the Raman line in natural diamond; the Raman spectra of later depositions did not show this large shift. The origin of the shift is not known. If the shift were due to a stress effect, the magnitude of the stress calculated on the basis of the stress dependence of the Raman shift in natural diamond would have to be above 3 x 10^9 Pa isostatic tensile stress.[12] Also seen in the spectrum is the Raman spectrum of a gem quality diamond and a mercury spectral lamp reference line used for calibration.

We were able to deposit continuous diamond films on Si single crystal substrates if, before deposition, the substrate was treated by rubbing the surface with 1 μm diamond paste or 1 μm diamond powder.[13] After rubbing, the substrate was cleaned, dried and placed in the substrate holder in the deposition chamber. The surface of the deposited continuous film, seen in Figure 4, possesses a significant surface roughness (about 2 μm peak-to-valley, 0.4 μm average roughness). With the realization of continuous film growth, characterization by other techniques could be achieved. Figure 5 shows Read XRD patterns from a diamond film and from diamond powder. Comparison of the two diffraction patterns shows a one-to-one correspondence between the diffraction lines in the two patterns thus identifying diamond in the film. Also note that there are no preferred orientation effects revealed in the diamond film XRD pattern. This has been found to be true for all diamond films examined by Read XRD.

One can also use two techniques of surface analysis to verify the existence of diamond in the films. One must remember, however, that these are surface techniques and the information comes from the top 2 to 3 nm of the film. The fine structure around the carbon KLL Auger peak for diamond and graphite is shown in Figure 6. The fine structure on the low energy side of the peak is clearly resolved in each case and compares well to the characteristic spectra published in the literature.[14] Figure 7 shows the differentiated electron energy loss spectra obtained from diamond film and graphite surfaces obtained using an electron beam of energy of 300 eV. Again, the fine structure is well resolved and the indicated peak positions agree well with those in the published literature.[14] A further confirmation of the presence of diamond phase in the films is obtained with Raman spectroscopy which, unlike AES and EELS, samples the bulk of the specimen. A Raman spectrum from one of the diamond films is shown in Figure 8. This spectrum differs significantly from that taken from an individual diamond particle. The characteristic peak is not shifted nearly as much and is broader in the case of the continuous film data. Near 1510 cm^{-1}, another Raman peak appears in the continuous film data which is due to non-sp^3 carbon bonds. Thus, the film appears to be a mixture of diamond and possibly diamond-like carbon (DLC). A significant background signal, attributed to specimen luminescence, extending over a large wavelength range, is also observed in the continuous film specimen. The luminescence signal, which has a peak near 738 nm (1.68 eV), is similar to a luminescence band in irradiated natural diamonds. This band is attributed to isolated lattice vacancies.

2.2 Depositions on different substrate materials

With the parameters for diamond film deposition established, attention turned toward examining the effects of substrate material, substrate orientation (in the case of single crystal Si) and substrate temperature on the morphology of the deposited films. Films were grown on Si single crystal plates on the (100) plane, 5° off the (100) plane, on the (111) plane, on fused silica, on boron doped single crystal Si, on polycrystalline silicon carbide, on polycrystalline silicon nitride and polycrystalline mullite.

Figure 9 displays SEM micrographs showing the morphologies of films deposited on Si, SiC and mullite and fused silica. All substrates in these cases had been rubbed with 1 μm diamond paste and cleaned before deposition. The deposition conditions were: substrate temperature, 800°C; gas composition, 99.5% H$_2$, 0.5% CH$_4$; deposition pressure of 5 x 10^3 Pa; and flow rate, 52 sccm. Deposition times were 60 hrs for the films on Si, 50 hrs for the films on SiC and mullite, and 45 hrs for the film on fused silica. Note that all films exhibit a similar morphology. Growth spirals are seen in the microstructure of the films. Grain size appears to be larger in the case of the film on SiC suggesting that the highest growth rates were achieved on these substrates.

The films on fused silica were cracked in many places; this was attributed to the large differences in the thermal expansion of the fused silica and the diamond film. These films had a water white appearance suggesting little optical absorption. Considerable optical scatter was present, however, due to the roughness of the film's surface.

In Figure 10, SEM micrographs are shown of films grown on Si_3N_4 and boron doped Si (111). The surfaces of the Si_3N_4 substrates had been polished with diamond compounds. One of the Si_3N_4 substrates was rubbed with diamond paste and cleaned prior to deposition; the other substrate was used as received. These films were deposited at substrate temperatures of 800°C, 99.5% H_2:0.5% CH_4 gas mixture, 5 x 10^3 Pa deposition pressure, 78 sccm flow rate for 68 hours. Continuous film growth was observed on both substrates with only a slight difference in the microstructure seen in the diamond rubbed and non-diamond rubbed Si_3N_4 substrates. It is observed that the scale of the microstructure is smaller for the films on Si_3N_4 than seen in films grown on other substrates.

In order to investigate whether continuous films could be grown without rubbing with diamond paste, a deposition was made on a Si substrate doped with concentration of 10^{20} boron/cc that was not rubbed with diamond paste. This is because it was believed that diamond could nucleate on boron. Deposited conditions were: substrate temperature, 800°C; gas composition, 99.5%H_2, 0.5%CH_4; gas pressure, 5 x 10^3 Pa; flow rate, 52 sccm; and deposition time, 52 hours. As can be seen in Figure 10, a continuous film was observed only on part of the specimen while other areas were only sparsely covered with individual particles (lower right micrograph Figure 10). The surface morphology on this film is similar to that seen in the previous micrographs.

From this series of micrographs it appears that the substrate exerts very little influence on the texture and morphology of the diamond films. The most noticeable variations were seen in the films grown on SiC, displaying larger grains and films on Si_3N_4 smaller grains. Deposition rates were about 0.1 μm/hr for all films, with the exception of SiC which had an apparently faster growth rate.

2.3 Effect of substrate temperature

A series of films were prepared to examine the effect of substrate temperature on film morphology. The deposition conditions were modified as follows: substrate material, Si rubbed with diamond paste and cleaned; flow rate, 52 sccm; substrate temperatures, 600°C, 650°C, 700°C, 750°C, 800°C, and 850°C; deposition time, 70 hrs. SEM micrographs of the films are shown in Figure 11 at low magnification and in Figure 12 at high magnification. A variation in microstructure is seen as the deposition temperature is changed. At low magnification, the variations in morphology show up as changes in the texture of the films. At the higher magnification, the details in the morphology changes can be observed. At 600°C, grains with a cube-like habit seem to predominate. At 650°C the microstructure is less well defined with triangular facets beginning to develop and mix with the cube-like habit. At 700°C the triangular faces predominate and the cube-like faces are absent. Growth spirals become evident on some of the faces. At 750°C the film surface displays well defined triangular faces, the crystal grains are well defined, and the growth spirals are more distinct. At 800°C the crystal grains are less well defined with secondary nucleation occurring as evidenced by the appearance of small grains growing between the larger grains. The growth spirals appear to have evolved into steps and ledges. At 850°C the secondary nucleation is much more pronounced with a greater loss of definition of the microstructure. No smooth facets are apparent.

It is evident from these micrographs that the substrate temperature strongly influences the microstructure of the deposited film. All these films displayed similar Read XRD patterns.

3. SUMMARY

The range of parameters in which diamond films can routinely and reproducibly be deposited has been established and verification of diamond films has been accomplished by a variety of analytic techniques. Continuous films, although rough, have been grown in a variety of substrates and at a range of substrate temperatures. Substrate material has little influence on the morphology of the film, but, substrate temperature affects the morphology of the deposited films strongly, apparently by promoting differing growth mechanisms at different temperatures.

4. REFERENCES

1. J.E. Field, editor, The Properties of Diamond, Academic Press, London (1979).
2. B.V. Derjaguin and D.V. Fedoseev, "The Synthesis of Diamond at Low Pressure," Sci. Amer. 102-109 (1975).
3. B.V. Spitsyn, L.L. Bouilov, and B.V. Derjaguin, "Vapor Growth of Diamond on Diamond and Other Surface," J. Cryst. Growth 52, 219-226 (1981).
4. S. Matsumoto, Y. Sato, M. Tsutsumi, and N. Setaka, J. Mater. Sci. 17, 3106-3112 (1982).
5. A. Sawabe and T. Inuzuka, "Growth of Diamond Films by Electron-Assisted Chemical Vapor Deposition," Appl. Phys. Lett. 46, 146-147 (1985).
6. A. Sawabe and T. Inuzuka, "Growth of Diamond Thin Films by Electron-Assisted Chemical Vapor Deposition and Their Characterization," Thin Solid Films, 137, 89-99 (1986).
7. K. Suzuki, A. Sawabe, H. Yasuda, and T. Inuzuka, "Growth of Diamond Thin Films by D.C. Plasma Chemical Vapor Deposition," Appl. Phys. Lett. 50, 728-729 (1987).
8. Z. Has and S. Mitura, "Nucleation of Allotropic Carbon in an External Electric Field," Thin Solid Films 128, 353-360 (1985).
9. Y. Saito, S. Matsuda, and S. Nigita, "Synthesis of Diamond by Decomposition of Methane in Microwave Plasma," J. Mater. Sci. Lett. 5, 565-568 (1986).

10. K. Kitahama, K. Hirata, H. Nakamatsu, S. Kawai, N. Fujimori, T. Imai, H. Yoshimo, and A. Doi, "Synthesis of Diamond by Laser-Induced Vapor Deposition," Appl. Phys. Lett. 49, 634-635 (1986).

11. R. A. Rudder, D. J. Vitkavage, and R. J. Markunas, "Growth of Diamond Films by Remote Plasma Enhanced Chemical Vapor Deposition," presented at the Second Annual Diamond Technolgy Initiative Seminar, Durham, NC, July 7-8, 1987.

12. A. Feldman, E. Farabaugh, Y. Sun, and E. Etz, "Diamond, A Potentially New Optical Coating Material," in Laser Induced Damage in Optical Materials: 1987, to be published.

13. A. Badzian, private communication.

14. P.G. Lurie and J.M. Wilson, "The Diamond Surface II, Secondary Election Emission," Surface Science 65, 476-498 (1977).

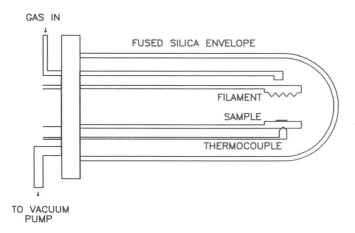

Figure 1. Apparatus for hot filament CVD of diamond films.

Figure 2. Diamond particles: (a) twinned icosahedron; (b) and (c) combination of {100} and {111} forms.

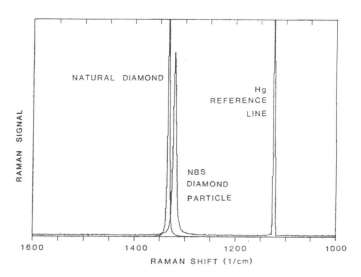

Figure 3. Micro-Raman spectrum of a diamond particle.

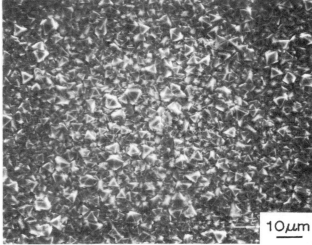

Figure 4. SEM micrograph of diamond film on (111) Si.

Figure 5. Read XRD patterns from diamond film on (111) Si substrate (a) and from diamond powder (b).

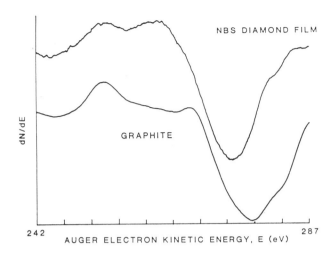

Figure 6.   Fine structure around carbon KLL Auger peak in diamond film and graphite.

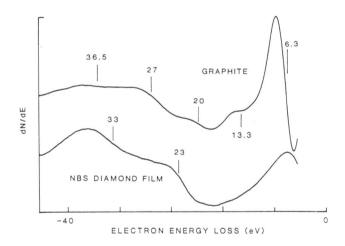

Figure 7.   Differentiated energy loss spectra from diamond film and graphite.

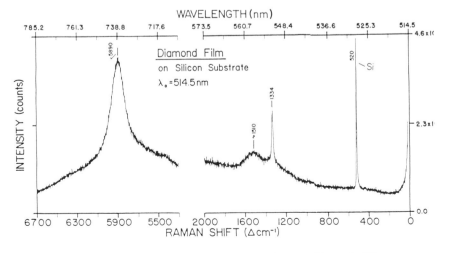

Figure 8.   Raman spectrum from continuous diamond film.

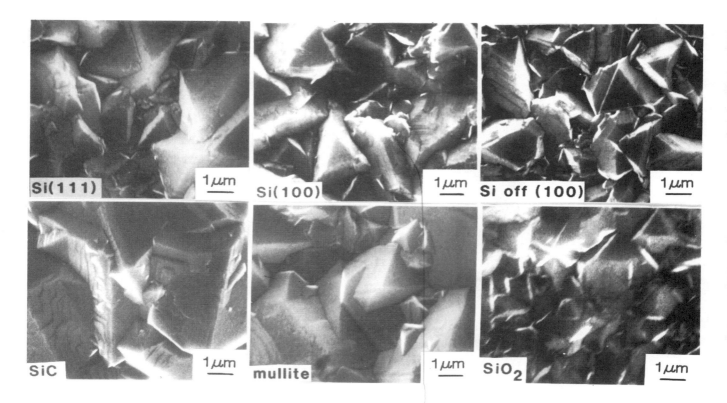

Figure 9. Diamond film morphology on Si, SiC, mullite and SiO_2.

Figure 10. Diamond films on Si_3N_4 and B-Si(111).

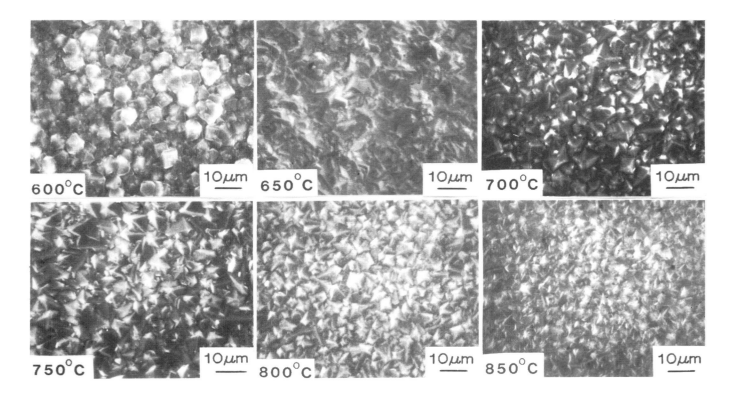

Figure 11. Lower magnification SEM micrographs of the effect of substrate temperature on diamond film morphology.

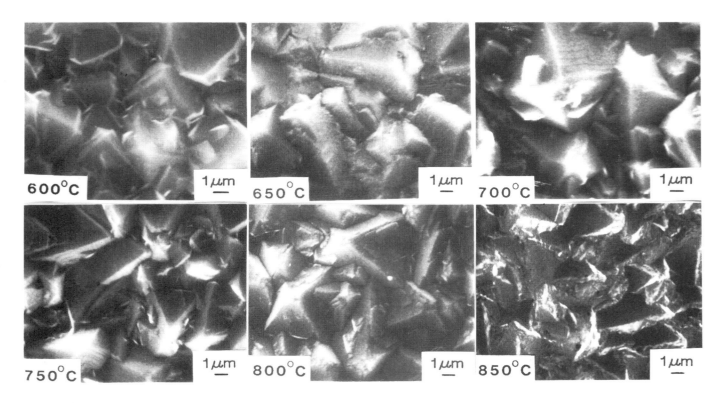

Figure 12. Higher magnification SEM micrographs of diamond films shown in Figure 11.

Diamond-like Carbon films by RF Plasma - enhanced Chemical Vapour

Deposition

R.S. Yalamanchi and G.K.M. Thutupalli

ISRO Satellite Centre,
Airport Road, Bangalore - 560 017, India

ABSTRACT

Diamond-like Carbon films synthesized by RF plasma-enhanced chemical vapour deposition were systematically characterized for their structural, optical and mechanical properties. The process parameters were optimized by the use of orthogonal tables to get the 'diamond-like' properties. DLC films were demonstrated to be efficient and durable antireflection coatings for IR devices. Plasma diagnostics were carried out by Langmuir probe method and Optical Emission Actionometry for Oxygen, Air and Water plasmas to give an understanding of the DLC etch mechanism.

1. INTRODUCTION

Diamond-like Carbon (DLC) thin films are of immense technological value, particularly in microelectronics and optics due to their exotic properties: extreme hardness, electronic insulating property, transparency in a wide spectral band, chemical inertness and dry etch compatibility. DLC films are being synthesized by a variety of techniques: DC and RF plasma deposition, Primary and Dual ion beam deposition, Microwave plasma deposition, Electron-assisted CVD and Hollow-cathode CVD. [1-4] The ion-based synthesis is believed to impart the 'diamond -like' properties to these films but the growth mechanism remains ambiguous. In this study, we optimized the film properties of particular concern to infrared viz., IR transparency and hardness. The optimized process conditions were employed to generate DLC films on silicon, Germanium, Quartz and other substrates for various analyses.

Plasma diagnostics were carried out by characterizing the Argon, Oxygen, Air and Water plasmas to understand the DLC etch mechanism under varying etch conditions.

2. DLC DEPOSITION AND PROCESS OPTIMIZATION

RF (13.56 MHz) plasma-enhanced chemical vapour deposition (PECVD) method was employed to synthesize DLC films on Si, Ge, Quartz and WC-Co. Prior to deposition, all the substrates were thoroughly cleaned in detergent and acetone baths ultrasonically followed by rinsing in 18 M ohm deionized water. DLC depositions were preceded by a sputter-etch step in argon discharges for periods of 5-20 minutes. Table.1 summarizes the experimental conditions.

TABLE 1. Experimental conditions

SUBSTRATES	Si, Ge, WC-Co, Fused Silica, Quartz
DISCHARGE FREQUENCY,f	13.56 MHz
CATHODE-ANODE SPACING,d	30-60 mm
RF POWER,P	0-100 W
SELF-BIAS VOLTAGE,V_B	100-2000 V
PROCESS GASES	C_2H_2, Ar, O_2, Air, C_6H_6, C_2H_6, H_2O
PRESSURE RANGE,p	$10^{-3}-10^{-1}$ Torr
SPUTTER ETCH TIME,t	5-20 min
SPUTTER ETCH POWER,P_E	0-50 W

The process parameters that could be varied to get the 'diamond-likeness' in the films were found to be the Self-bias Voltage (V_B), Acetylene pressure (P), Argon-to-Acetylene transition time (t_T) and Cathode-to-Anode spacing (d). The output functions monitored were : IR transparency (or Absorption), Microhardness (H_V) and Refractice Index (n). The input parameters were varied over three level settings after an exhaustive one-dimensional search. These fortutiously fit the orthogonal matrix $L_9 3^4$ ($L_x z^y$, where y is the number of input parameters, z is the number of level settings and x is the number of experimental runs to complete the matrix). The output function averages (arithmatic means) for each level setting and for each input parameter were determined. For each level of self-bias voltage, the orthogonal property of the matrix randomizes the settings for the other three input parameters, so that the effect of each of these parameters tends to cancel out.[5] Continuing in the same vein, the parameters more critical towards the 'diamond-likeness' were determined to be the self-bias voltage and the Acetylene pressure. Cathode-to-Anode spacing had the least impact of all the parameters on the film properties.

Thus the recipe concocted for the DLC films yields the RF self-bias voltage of 980V, Acetylene pressure of 20 mTorr, Argon-to-Acetylene transition time of 60 seconds and a cathode-to-Anode spacing of 60mm. The process was further optimized to include the substrate temperature (which depends on the RF power to the cathode) and process gas admixtures by including factor interaction columns in the orthogonal design. The refined experimental design yielded a RF cathode coolant temperature of $15^{\circ}C$ for a power density of $1.0 W/Cm^2$ and the addition of a plasma stabilizing gas (hydrocarbons like C_6H_6, C_2H_6). The effect of gas additives is depicted in Table.2.

TABLE 2. Effect of gas Additives on DLC film synthesis

Additive	Purpose	Example Active gas-Additive gas
Rare gas	Stabilize plasma	
	Sputter-etch loosely bound ad-atoms	C_2H_2-Ar
	Control over deposition rate	C_2H_2-He
	Increased ion bombardment due to Penning ionization	
Hydrocarbon	Enrichment of C/H Species in the plasma	C_2H_2-C_6H_6
	Stabilize plasma	
Etchant	Reactive ion etching of the layer	C_2H_2-O_2 C_2H_2-Air
		C_2H_2-H_2O

3. FILM CHARACTERIZATION

DLC films deposited on Si wafers and floated on Transmission Electron Microscope grids were predominantly amorphous in nature. No indication of microcrystallinity was observed by varying the deposition parameters. IR absorption studies on DLC coated Si and Ge in the 4000 Cm^{-1} - 200 Cm^{-1} range showed C-H (SP^1, SP^2, and SP^3) stretch bands at 3300, 3000 and 2920Cm^{01}; C-H_2 deformation and C-H out-of-plane bending bands in the range of 1500-700 Cm^{-1}.We found a marked lowering of C-H absorption due to increased substrate temperature and this phenomenon has been effectively used to optimize the IR transparency in the IR bands for imaging applications. Spectrophotometric analyses and closed-loop iterative methodology yielded refractive indices in the range 1.8-2.15 and extinction coefficients in the range 8×10^{-2} - 6×10^{-2} with varying process parameters. Ellipsometric studies at three different wavelengths confirmed these measurements.

Tauc analysis, viz., $(AE)^{1/2}$ versus E of DLC films, where A is the absorbance and E, the photon energy showed a decrease in bandgap with an increase in RF power density, in confirmation of the earlier results.[6,7] This can be related to the loss of hydrogen as ion bombardment increases. The results fit the Mott-Davis model for amorphous materials and yield the bandgap in the range 0.8 - 1.7 eV.

DLC films were typically characterized to be of high hardness and scratch resistant, making them attractive as protective coatings. A rough estimate of scratch resistance was obtained by testing with a razor edge. A transition from easily scratched 'soft' to unscratched 'hard' was observed as the ion bombardment increased. High internal stresses made thick DLC films to delaminate on quartz substrates, though they were adherent to Si and Ge. Microhardness measurements were made on DLC deposited Si, Ge, WC-Co at different loads and the error due to substrate was considered in the measured hardness numbers of 1500-3200 under different process conditions.

4. DLC AS EFFICIENT COATING FOR IR SENSOR OPTICS

DLC acts as a durable and efficient anti-reflection coating material for IR sensor optics in 3-5, 8-12 and 14-16 micron bands. Various configurations : DLC/IR optic/DLC ; DLC/IR optic/High efficiency coating, were experimentally realized with the optimized process conditions and these pass the stringent MIL-specifications. Some configurations are listed in Table.3.

TABLE 3. DLC based AR configurations

SYSTEM	SPECTRAL CHARACTERISTICS (T % Avg.)			REMARKS
	3-5microns	8-12microns	14-16microns	
DLC/Si/DLC	90	-	-	IR Optics pass the following environmental & durability tests:
DLC/Ge/DLC	90	88	80	- Adherence acc.MIL-M-13508B - Severe abrasion acc.MIL-C-675A - Salt spray acc.MIL-C-675A
DLC/Si/SiO	92	-	-	- Thermal cycling acc.MIL-M-13508B - Humidity/240h acc.MIL-C-48497

DLC/Si/Al O
     x y
	92	-	-	- Thermal Shock - Immersion in boiling water followed by immersion in LN$_2$
				- Chemical attack : 60 min. in HCl/H SO /HF or NaOH (0.1N)
				        2 4

DLC/Ge/ZnS	92	92	85	

5. REACTIVE ION ETCHING (RIE) STUDIES

Reactive ion etch studies on DLC films were carried out to tune the film thickness and to test the dry etch compatibility for microelectronic applications.[9]

5.1 Plasma Diognostics:

Plasma diagnostic experiments: Optical emission actinometry and Langmuir probe technique were employed to characterize Argon, Oxygen, Air and Water plasmas. The emission spectra were recorded through a fibre-optic light guide which monitors the plasma glow through a window in the plasma reactor, and a Rofin spectralyzer/Monochromator set-up. Typically for the case of O_2 plasma, the emission lines of interest were 845 nm and 778 nm from O atoms and a 560 nm emission line from O_2^+. A small amount of inert gas (Argon) was bled into the system and all the emission lines were normalized with respect to the emission lines for Argon (750nm) to account for the changing excitation efficiency under different plasma conditions. Emission intensities thus normalized give a fair approximation of the concentrations of the corresponding species in their ground states, if the excited states lie in the same energy range, as has been in our case and if excitation occurs directly by electron impact from the ground state.[10].

Langmuir probe measurements supplemented by the above studies give the electron densities in the range of 10^{10} cm^{-3} and electron energies to be 1.8 - 10eV for the present system which are in good agreement with the results obtained by other groups.[11].

Fig.1 illustrates the O atom concentration mapped out from the emission intensities in the plasma diagnostic experiments. It can be clearly seen that the O atom concentrations are higher in Water plasmas than in Oxygen plasmas. This can be attributed to the unimolecular decomposition of H_2O^+ formed by electrom impact in the bulk of the plasma. As a result of the multiple-electron impact induced dissociative steps, the number density of atomic oxygen increases as is clearly evident in our case.

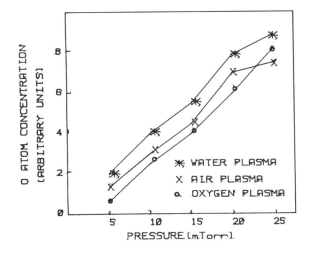

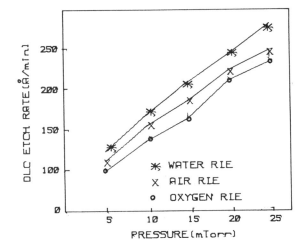

Fig.1 Plasma Diagnoistic study

Fig.2 DLC Etch rate in different plasmas

5.2 DLC Etch Mechanism

The DLC etch rate also undergoes a rather dramatic increase with the induction of H_2O species into the plasma. Fig.2 illustrates this phenomenon. DLC etch mechanism can beunderstood to proceed by the formation of volatile products like CO, CO_2, H_2O by the highly reactive species in the plasma. A closed-loop plasma CVD-RIE illustrates the thickness tuning of DLC with intermediate breaks to test the thickness measurements shown by flat regions.

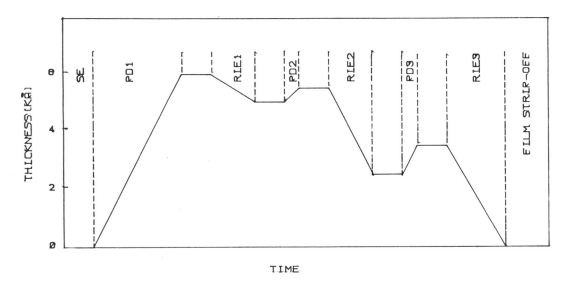

Fig.3 Schematic of PD-RIE cycle of DLC on Ge

6. CONCLUSIONS

DLC deposition process optimization by factorial experimental design yielded the diamond like properties for IR devices applications to be Self-bias voltage of 980V, Acetylene pressure of 30 m Torr, Argon-to-Acetylene transition time of 60 seconds and a Cathode-to Anode spacing of 60 mm with an admixture of a plasma stabilizing hydro-carbon gas. AR designs realized over different spectral bands demonstrate the suitability of DLC. Reactive ion etching studies indicate that Water plasmas might turn out to be better dry-etchants compared to Oxygen plasmas.

7. ACKNOWLEDGEMENTS

The authors wish to thank N. Pant, Director and T.K. Alex, Head Sensor Systems Division for the kind encouragement throughout the course of this work.

8. REFERENCES

1. A. Bubenzer, B. Dischler, G. Brandt and P. Koidl, "Role of hard carbon in the field of infrared coating materials," Opt. Engg. 23(2), 153-156 (1984)

2. J.C. Angus, P. Koidl and S. Domitz, "Carbon Thin Films", in Plasma Deposited Thin Films, J. Mort and F. Jansen, Ed., PP.89-127, CRC Press, Florida. (1986).

3. M.J. Mirtich, D.M. Swec, J.C. Angus, "Dual Ion beam Deposition of Carbon films with Diamond-like Properties," TM-83743, NASA Lewis Research Center, Cleveland (1984).

4. B. SIngh, Y. Arie, A.W. Levine and O.R. Mesker, "Effects of filament and reactor wall materials in low-pressure chemical vapour deposition synthesis of diamond", Appl. Physics. Lett. 52 (6), 451-452 (1988).

5. W.J. Diamond, Practical Experimental Designs, Van Nostrand Reinhold Co., New York (1981).

6. H. Tsai, D.B. Bogy, "Characterization of diamond-like carbon films and their application as overcoats on thin film magnetic recording", Critical Review, J. Vac. Sci. Technol. A 5(6), 3287-3312 (1987)

7. C.J. Robinson, M.G. Samant, J. Stohr, V.S. Speriosu, C.R. Guarnieri and J.J. Cuomo, "Structural studies of hydrogenated amorphous carbon infrared coatings", Mat. Res. Soc. Symp.Proc., 90 (1987)

8. J.A. Savage, Infrared Optical Materials and their Antireflection Coatings, Adam Hilger Ltd, Bristol (1985).

9. N.G. Einspruch, D.M. Brown, VLSI Electronics-Microstructure Science Vol.8, Academic Press Inc. Florida (1984).

10. R.A. Gottscho and V.M. Donnelly, "Optical emission actinometry and spectral line shapes in rf glow discharges", J. Appl. Phys., 56, 245-250 (1984).

11. D.A.O. Hope, T.I. Cox and V.G.I. Deshmukh, "Langmuir probe and optical emission spectroscopic studies of Ar and O_2 plasmas," Vacuum, 37, 275-277 (1987).

A Study of the Doping Process in Diamond
by Boron Implantation

G.S. Sandhu, W.K. Chu, M.L. Swanson, and J.F. Prins*
University of North Carolina, Dept. of Physics & Astronomy
Chapel Hill, North Carolina 27599-3255

ABSTRACT

Type IIa diamond crystals were implanted with boron ions with or without prior carbon ion implantation. The samples were kept at liquid nitrogen temperature during both implantation steps. A strong near-edge optical absorption band appeared after implantation, and partially recovered during annealing at 800 °C. For the highest B implantation fluence, optical absorption peaks at 2800 to 3000 cm^{-1} were observed that were in the same vicinity as the absorption peaks attributed to substitutional boron atoms in natural p-type diamond. Electrical measurements for three of the samples demonstrated well-defined activation energies that could be associated with hopping conduction and/or activation of B dopant atoms. This work shows that p-type doping in diamond by boron ion implantation is feasible, using a suitable combination of low temperature implantation and subsequent annealing.

*Permanent address: University of the Witwatersrand, Johannesburg, South Africa.

1. INTRODUCTION

Recent successes in diamond film growth and the possibility of diamond-based microelectronics have triggered a flurry of activity in diamond research.[1] Ion implantation seems to be one of the most promising ways of introducing dopant atoms into diamonds. Attempts to dope diamond by ion implantation have had a long history.[2] Unfortunately, due to the scarcity of diamond samples, most of the early experiments were seldom reproduced. Many of the original claims were intuitively sound but not empirically verified.

Vavilov et al., established at an early stage that electrical conductivity can be obtained by ion implantation.[2,3] They reported that boron or aluminum implanted diamonds were p-type while lithium, carbon and phosphorous implanted diamonds were n-type. Due to the difficulty of making ohmic contacts in the early days, the A.C. conductivities were measured by the microwave method. They also claimed that Hall measurements of boron-implanted diamonds showed a p-type behavior[4] and that p-n junctions could be obtained by boron plus phosphorous implantations.[5] However, complications with interpretation do exist because the intrinsic radiation defects in diamond are also electrically active and may even cause impurity band conduction by means of the hopping mechanism.[6-8]

Diamond is a metastable form of carbon. Thermodynamically, annealing of ion-damaged diamond will thus favor graphite formation unless the kinetics of the process can prevent the transition. Hot implantation was proposed in the hope that dynamic annealing would occur during the implantation process.[9,10] Although graphitization can be prevented in this manner, a dense dislocation network results caused by the interaction of the migrating point defects.[11]

At lower temperatures, where this dynamic annealing does not occur, implantation causes volume expansion. Maby et al.,[12] reported expansion of diamond after boron ion implantation at room temperature and concluded that this was caused by the reduction of material density needed for amorphous carbon formation in diamond. Volume expansion of diamond measured during ion implantation at different temperatures was studied by Prins et al.[13,14] They observed that no volume expansion could be detected for implantation at liquid nitrogen temperature up to a threshold dose of 6×10^{15} ions/cm^2.[14] This low temperature result contrasts sharply from ion implantation carried out above room temperature when the expansion occurred already at the onset of ion implantation.[13] The latter volume expansion was attributed to the outdiffusion of carbon interstitials from the ion damaged region, leaving behind a high density of immobile vacancies. In contrast, for implantation at liquid nitrogen temperature, the carbon interstitials as well as the vacancies in the vicinity of the damaged area are immobile.

The above observations are crucial for implantation doping and led Prins[15] to devise a model and strategy which demonstrated that controlled doping of diamond by means of ion implantation can be achieved. According to the recipe, ion implantation should be carried out at a low enough target temperature (for example, liquid nitrogen temperature) to inhibit diffusional motion of the point defects created during the process. This is followed by a suitable annealing cycle which then enhances the probability of dopant-vacancy combination owing to the large defect density. The efficiency of this annealing process should improve when increasing the initial point defect density "frozen" in during cold implantation. This density can be increased relative to the implanted dopant atom density by implantation of, for example, carbon ions.[15] However, the density should be below the threshold value at which the layer will expand to form graphite.[14] In a follow-up paper, Prins[16] also demonstrated that the annealing efficiency can be greatly improved by heating the diamond rapidly to a higher temperature than used for his initial studies. Obviously, the efficiency of this doping process can be improved by optimizing the implantation parameters and the subsequent annealing cycle.

Preliminary experimental results which were obtained when implanting diamond with boron plus carbon ions at low temperature according to the recipe discussed above, are presented in this paper. This constitutes a report on the ongoing progress of our research on this subject.

2. EXPERIMENTAL

Type II-A insulating [110] single crystal diamonds of $4 \times 4 \times 1 mm^3$ (purchased from Dubbeldee Diamond Company) were cleaned in hot chromic acid prior to implantation. The boron implants were done with and without prior carbon implantation at liquid nitrogen temperature. When the double species (C and B) were implanted, the diamond samples were kept cold and inside the implanter without breaking the vacuum. The energy and dose of the implants are summarized in Table I. The implantation energies were selected to match the range of boron ions with that of the vacancy distribution produced by the previous carbon

TABLE I

SAMPLE	IMPLANTED ION	ENERGY (keV)	DOSE (atoms/cm^2)
1	B	65	1×10^{14}
2	C	110	1×10^{14}
followed by	B	65	1×10^{14}
3	C	200	3×10^{15}
followed by	B	120	1×10^{15}
4	C	200	3×10^{14}
followed by	B	120	3×10^{15}
5	C	200	3×10^{14}
followed by	B	120	1×10^{15}

implantation. This was achieved with the help of computer simulations using the TRIM-88 program.[17] The probability of boron atoms occupying vacant sites in the lattice is expected to be enhanced during ion irradiation and post implant annealing if most of the boron atoms are in the vacancy-rich region. All the implantations were done at liquid nitrogen temperature and the samples were then brought slowly to room temperature. After implantation, some of the diamonds were subjected to isochronal annealing in vacuum (5×10^{-6} Torr). Rapid thermal annealing (RTA) was also done.

Optical measurements were made at room temperature in the transmission mode using a Nicolet 20DXB FT-IR Spectrophotometer. The optical spectra of all the diamonds were measured before and after ion implantation and after subsequent annealing.

On sample 4, electrically active regions which penetrated through to the implanted layer were made by laser irradiation of the diamond surface to produce a heavily damaged region, and contacts were then made with spring loaded tungsten probes. The current injection into the diamond between two contacts was typically between -100 and +100 nA when the voltage readout covered the region between -5 volts and +5 volts. The resistance R was obtained from the slope of the V vs I trace near the origin. The sample was mounted on a heater and the measurements were made at temperature between 36°C and 300°C.

The samples 3 and 5 were annealed at 900°C for 1 hour and rapid thermal annealed (RTA) at 1100°C for 2 minutes. Optical measurements were then made on these samples and contacts for electrical measurements made by overdoping the contact surfaces.[18] The resistance of these samples was then measured as a function of temperature between 20°C and 300°C.

3. RESULTS

The results of the optical measurements are shown in Figs. 1 and 2. Figure 1 shows the absorption spectra in the UV-VIS region for sample 1 (implanted with 65 keV B to a dose of 10^{14} B/cm^2) after annealing at different temperatures for 45 minutes. It shows the absorption band (GR-I)[19,20] caused by implantation damage. The gradual annealing of the radiation damage is indicated by the decreasing absorption of the GR-I band and a more dramatic upward movement of the absorption edge at 220nm. It is obvious that the radiation damage was not completely removed even after the 800°C anneal. The step at 800 nm is an artifact due to source and filter change in the instrument.

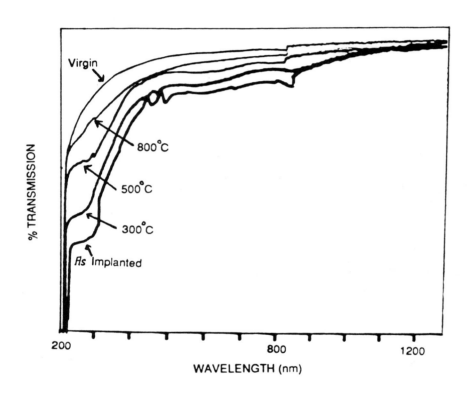

Fig. 1 Optical transmission spectra of diamond implanted with 65 keV boron ions at 77 K to a dose of 1×10^{14} ion per cm^2 followed by annealing at different temperatures for 45 minutes.

Absorption spectra for sample 4 in the far infrared region are shown in Fig. 2. For comparison, the spectrum obtained from a natural semiconducting diamond containing boron (type IIb) is also shown. After implantation, sample 4 showed two small absorption peaks at 2957 cm^{-1} and 2925 cm^{-1} (o.365 and 0.361 eV), which were not present before the implantation. These peaks correspond closely to the position of one of the boron peaks at 0.363 eV observed in the natural semiconducting diamond and it is thus tempting to conclude that the appearance of these peaks indicates the presence of substitutional or near-substitutional boron in the diamond lattice.[21] However, this conclusion must be approached with caution because the dominant peaks at 0.345 eV and 0.305 eV which are always present in natural semiconducting diamond containing boron did not appear after implantation (see Fig. 2). The relative magnitude of the two peaks which are caused by ion implantation into sample 4 was small because

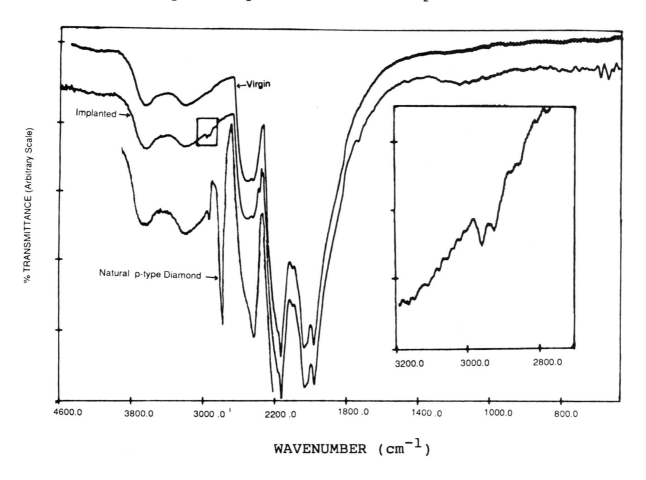

Fig. 2. Absorption spectra for sample 4 (see Table 1) before and after implantation, compared with a spectrum for type IIb diamond. The inset shows the absorption peaks at 2957 cm^{-1} and 2925 cm^{-1} for the implanted sample 4.

the thickness of the implanted layer was a minute fraction of the total thickness traversed by the probing beam. It was concluded that in the first two cases of Table I, the boron ion dose could have been too small to cause observable peaks at the same position as seen in sample 4. Unfortunately, samples 3 and 5 already displayed the same peaks before ion implantation and no conclusion could thus be drawn from these measurements. If the peaks were, in these cases, enhanced by the implanted boron, the effect was too small to be measured. It is interesting to note that these peaks can be present in unimplanted, natural diamonds which have been classified as insulating (type IIa).

Because the peaks appeared in sample 4 directly after ion implantation, it was decided to characterize the implanted layer electrically without annealing the diamond. A plot of resistance vs 1/T, given in Fig. 3, yields an activation energy of 0.28 eV, which is

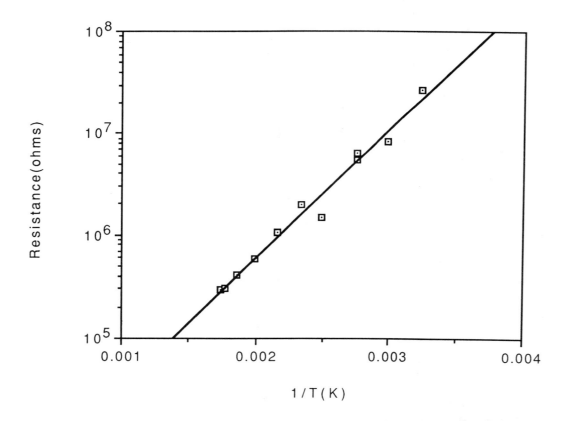

Fig. 3 Electrical resistance versus reciprocal temperature for sample 4 (see Table I) before annealing.

24% below the recognized energy level of substitutional boron in diamond (0.37 eV). This low value, in conjunction with the low resistivities measured, indicates that p-type conductivity by hole activation from boron-acceptors is not the dominating process. The electrical behavior is reminiscent of the initial results which were published by Vavilov et al.[2-5] It would seem logical to conclude that radiation damage rather than substitutional boron is dominating the conduction. This conclusion received further support when this implanted layer turned graphitic after annealing this diamond at 1200°C for 1 hour. Any substitutional boron would have been heavily compensated by the large number of vacancies which act as donors.[22] Thus, if the infra-red absorption peaks which appeared after implanting sample 4 were caused by the presence of substitutional boron, it must be heavily compensated which, in turn, may explain the absence of the other boron peaks which should be there. If this is true, it would mean that samples 3 and 5 already contained compensated substitutional boron before ion implantation. This correlates with the fact that thermally activated conductivity can be measured at high temperatures in type IIa diamonds and that the activation energies seem to be of the correct order of magnitude to explain conduction via hole generation from the charged donors which are compensating boron acceptors.[18]

Samples 3 and 5 were annealed at 900°C for 1 hr and at 1100°C for 2 min using RTA. Neither of the implanted layers graphitized during these anneals. This is a significant result in view of the fact that sample 3 contained more radiation damage (a larger total ion dose) than sample 4 which, as discussed above, graphitized when it was slowly heated and annealed at 1200°C. In the latter case, the residual vacancy density (i.e., the number of vacancies that did not recombine with interstitials) was high enough to induce graphitization. Another noteworthy aspect is that the two small infra-red absorption peaks (at 0.365 and 0.361 eV) which were present in those diamonds and correspond to the peaks which arose in sample 4 after ion implantation, disappeared after RTA. The fact that these peaks were also present in samples 3 and 5 before ion implantation shows that RTA also caused annealing in virgin, unimplanted diamond.

The conductivity vs 1/T plots for samples 3 and 5 are shown in Fig. 4. Sample 3 showed higher conductivity and a lower activation energy of 0.13 eV at room temperature, whereas sample 5, which received a lower total dose, had lower conductivity and a higher activation energy of 0.2 eV at room temperature which increased to 0.38 eV above 250°C. The four orders of magnitude difference in the conductivities of the two samples may be explained by the higher carbon dose received by sample 3. This will create more vacancies to trap boron atoms in substitutional sites during annealing, resulting

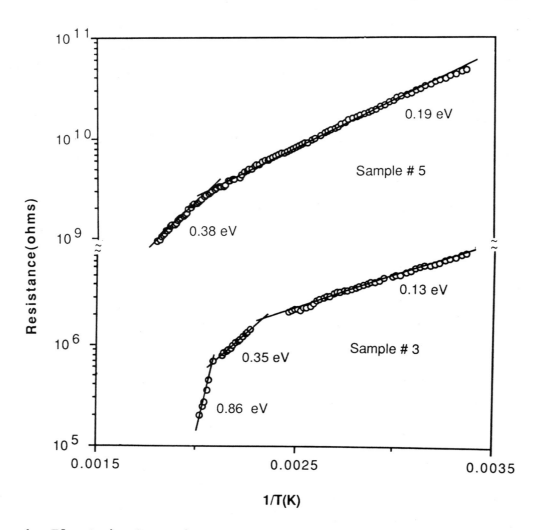

Fig. 4 Electrical resistance versus reciprocal temperature for samples 3 and 5 (see Table I) after 1100°C RTA.

in more efficient doping. On the other hand, the excessive damage produced by these extra carbon ions may also result in higher conductivity after annealing. However, the increase in the slope around 200°C in the case of sample 3 is most probably caused by the interplay between boron acceptors and radiation damage energy levels in diamond, and suggests that there is a significant amount of substitutional boron present in the sample. We expect that the same phenomenon would be observed in the case of sample 5 at higher temperatures.

4. CONCLUSIONS

In summary, doping of diamond by boron ion implantation at low temperature was studied. The diamonds were pre-implanted with carbon in order to increase the doping efficiency. Various post implantation annealing procedures were employed to drive the boron atoms into substitutional sites. These implanted samples were characterized by optical and electrical techniques. It was established that absorption peaks observed in the infrared spectra, which were thought to relate to substitutional boron, could be removed using RTA. The electrical measurements showed features similar to those reported by Prins[16] and indicated the presence of substitutional boron in the implanted samples. The various experimental parameters were by no means optimized because of the limited number of available samples. Nevertheless, the results will be of great assistance in planning future experiments.

5. ACKNOWLEDGEMENTS

This work is supported by the Office of Naval Research contract N00014-87-K-0243. Ion implantation was done in the microelectronics laboratory (North Carolina State University site) of the Microelectronics Center of North Carolina, and optical measurements were done at Research Triangle Institute of North Carolina. We thank M.W. Geis of MIT Lincoln Laboratory, Lexington, MA. for making the electrical measurements for sample 4.

6. REFERENCES

1. For example, the Third Annual Diamond Technology Initiative Symposium, July 12-14, 1988, in Arlington, VA has attracted over 60 oral presentations.

2. V.S. Vavilov, M.I. Guseva, E.A. Konorova, V.V. Krasnopevtsev, V.F. Sergienko, and V.V. Tutov, Sov. Phys. Solid State 8, 1560 (1966)

3. V.S. Vavilov, M.I. Guseva, E.A. Konorova and V.F. Sergienko, Sov. Phys. Semiconductor 4, 6 (1970).

4. V.S. Vavilov, M.I. Guseva, E.A. Konorova, and V.F. Sergienko, Sov. Phys. - Semiconductors 4, 12 (1970).

5. V.S. Vavilov, M.A. Gukasyan, M.I. Guseva, E.A. Konorova, and V.F. Sergienko, Sov. Phys. - Doklady 16, 856 (1972).

6. J.J. Hauser and J.R. Patel, Sol. State Comm. 18, 789 (1976).

7. J.J. Hauser, J.R. Patel, and J.W. Rodgers, Appl. Phys. Lett. 30, 129 (1977).

8. R. Kalish, T. Bernstein, B. Shapiro, and A. Talmi, Rad. Eff. 52, 153 (1980).

9. R.S. Nelson, J.A. Hudson, and D.J. Mazey, British patent No 1 476 313, published 10 June 1977.

10. R.S. Nelson, J.A. Hudson, and D.J. Mazey, British patent No 1 599 668, published 7 October 1981.

11. R.S. Nelson, J.A. Hudson, D.J. Mazey, and R.C. Piller, Proc. R. Soc. London A386, 211 (1983).

12. E.W. Maby, C.W. Magee, and J.H. Morewood, Appl. Phys. Lett. 39, 157 (1981).

13. J.F. Prins, T.E. Derry, and J.P.F. Sellschop, Phys. Rev. B34, 8870 (1986).

14. J.F. Prins, T.E. Derry, and J.P. F. Sellschop, Nucl. Instr. & Meth. B18, 261 (1987).

15. J.F. Prins, Phys. Rev. B38, (to be published 1988).

16. J.F. Prins, IBA Conference, April 1988, Johannesburg, South Africa(to be published in Nucl. Instrum. & Methods)

17. J.F. Ziegler, J.P. Biersack, and U. Littmark, The stopping Range of Ions in Solids, Pergamon, New York, 1985.

18. J.F. Prins, to be published.

19. E.C. Lightowlers and A.T. Collins(1976). Diamond Research 1976 (Suppl. Ind. Diam. Rev.) pp 14-21.

20. H.B. Dyer and L. du Preez, J. Chem. Phys., 42, 1898 (1965).

21. G.S. Sandhu, M.L. Swanson, and W.K. Chu, IBMM '88, June 12-17, 1988, Tokyo, Japan.

22. H.B. Dyer and P. Ferdinando, Br. J. Appl. Phys. 17, 419 (1966).

SESSION 3

Characterization I

Chair
John A. Woollam
University of Nebraska/Lincoln

Invited Paper

Bonding, interfacial effects and adhesion in DLC.

A.Grill*, B.S.Meyerson and V.Patel

IBM Research Division, T.J.Watson Research Center, Yorktown Heights,
N.Y. 10598.

ABSTRACT

Metastable, amorphous carbon films, also referred to as diamond-like carbon (DLC), are prepared in a large variety of deposition conditions. Depending upon the conditions, DLC films may contain large amounts of hydrogen, and are referred to as hydrogenated amorphous carbon. The properties of DLC are strongly dependent on the preparation conditions and upon the amount of hydrogen incorporated in the film. Due to its extreme hardness, DLC can be used as a wear protective coating, while its chemical inertness to acids and alkali's make it suitable for protection against chemical attack.

The optical transparency of the DLC over a large region of the spectrum makes it useful as a protective coating of optical components. However, hydrogenated DLC is usually also characterized by very high compressive stresses. Its application as a protective coating therefore requires strong adhesive bonds to the coated surface. The paper reviews structural and optical properties of DLC films and means of improving their adhesion to metallic surfaces.

1.INTRODUCTION

Diamond-like carbon, or DLC, films are metastable amorphous materials, which may include a microcrystalline phase. Diamond layers, now fabricated employing plasma based deposition methods, are polycrystalline materials, with crystallites up to tens of microns in size, having the diamond structure. Since first deposited by Aisenberg and Chabot in 1971[1], the DLC films have been prepared by a variety of methods including DC glow discharge, RF plasma (PECVD), sputtering, and ion beam deposition, from a variety of carbon bearing, solid or gaseous source materials[2-7]. All the deposition processes of DLC are characterized by the interaction of energetic ions with the surface of the growing film. Depending on the precursor materials, many of these films contain a significant amount of hydrogen. For varied deposition methods, the hydrogen content in DLC films was determined to be between less than 10%[8] to 50%[9]. The metastable structure of DLC films most likely originates from the thermal and pressure spikes caused by impinging energetic ions on the surface [10]. Thermal spikes of 3283 K and pressures of 1.3×10^{10} Pa (1.2×10^5 atm), for about 7×10^{-11}s have been calculated for impinging 100eV ions[11]. The duration of the spikes is much longer than the vibration period of 2.6×10^{-14}s for diamond. The formed metastable structure is conserved due to extremely high quenching rates of the thermal spikes.

DLC films are characterized by extreme hardness, measured to be in the range 3000-9000 Kg/mm^2 [12], a generally low friction coefficient between 0.01[13] and 0.28[14], and very high internal stresses[4 7 13 15]. The films typically have high optical transparency over a wide spectral range, high electrical resistivity, and chemical inertness to both acids and alkalis.

DLC films may contain sp^2, sp^3 and even sp^1 coordinated carbon atoms in a disordered network. The ratio between the carbon atoms in the different coordinations is to a great extent determined by the hydrogen content of the films. Hydrogen is also important if one is to obtain a wide optical gap (E_{opt}) and high resistivity, as it passivates the dangling bonds in the amorphous structure[11]. Inevitably, if one wishes to understand the physical basis for the properties of DLC, such as hardness, resistance to chemical attack, or optical transparency, one must characterize the basic atomic structure of the material. Of obvious importance is the ratio of carbon atoms in each bonding coordination, the hydrogen content of the film and its distribution in the bulk of the film.

2.BONDING

High resolution ^{13}C NMR spectroscopy employing proton decoupling has been used to determine the relative concentrations of sp[2] and sp[3] hybridized carbon, as well as the local atomic environment of carbon in each state. ^{13}C NMR spectra can be obtained at a center frequency of 48.29 MHz, using a 180 MHz proton decoupling signal. Adequate peak separation can be obtained using magic angle spinning at 3.5 kHz[16]. Obtaining spectra both with and without the use of proton decoupling enables the identification of carbon atoms with bound hydrogen, because the C-H spin interaction causes the carbon peak to be broadened into the baseline of the spectrum, leaving only the signal from carbon atoms with no bound hydrogen[17].

Typical results obtained for DLC films prepared by PECVD from acetylene are presented in Fig.1 which shows the sp[2] and sp[3] peaks obtained with and without proton decoupling[15]. The results indicated a ratio of sp[2]:sp[3] hybridized carbon of 1.6, and that virtually all sp[3] carbon atoms are in fact bound to one or more hydrogen atoms[15]. A significant amount of hydrogen was also found to be bound to sp[2] carbon, as indicated by the large decrease of the sp[2] peak in Fig.1.B.

Similar results have been found by Jarman and Ray[18] for PECVD films deposited from methane and by Jansen et.al.[19],[20] for ion beam sputter deposited films and PECVD films. Jarman found that hydrogen is bound to both sp[2] and sp[3] carbon, though primarily to sp[3] carbon, while Jansen found that the sp[2] bonds in ion beam sputter deposited films remain to a large extent unsaturated up to 35 at.% hydrogen.

Quantification of the sp[3], sp[2] and sp[1] hybridization of the DLC films has also been obtained by deconvoluting the vibrational bands in the IR absorption spectra. A relative concentration of sp[3]:sp[2]:sp[1] = 68:30:2 has been obtained for DLC films deposited from benzene[21]. After annealing above 300 C the sp[3] carbon converts to sp[2], reaching 100% sp[2] at 600 C[21]. IR spectroscopy also indicated that in RF plasma deposited DLC films the majority of hydrogen is bound to the sp[3] carbon[22]. These results are the inverse of the NMR results presented above[15],[18] indicating how system dependent such properties can be. After annealing 1 hour at 500 C the sp[3] bound carbon evolves its hydrogen, forming sp[2] bonds, but without significant loss in IR transparency. After annealing at 700 C all hydrogen is removed from the films, resulting in the loss of IR transparency[22].

3.INTERFACIAL EFFECTS

For many applications of DLC, layers of order 100 nm thickness or less have to be used, and special care must be taken in defining the properties and and their distribution through film thickness. Characterization methods such as optical measurements[3][4][23] and NMR[15][18-20] typically measure spatially averaged bulk properties of the films, although in reality film anisotropy can result in substantial variations of the properties of interest as a function of film thickness. The existence of anisotropy can be explicitly demonstrated by hydrogen profiling and XPS measurements of DLC films.

The energy of the C_{1s} XPS peak is determined by the bound state of the carbon atom, i.e. incorporated in an adsorbed hydrocarbon species, complexed as a carbide, or in its graphitic or diamond state[24],[25]. There is a separation of at least 1eV between these different carbon peaks, and Mori and Namba[24] used XPS techniques to argue that their carbon films had a diamond structure. Fig.2 presents XPS spectra obtained at both normal and grazing incidence on a 40 nm thick film deposited by PECVD from an acetylene source[15]. A shift of the binding energy towards higher values is observed for spectra taken close to the surface normal relative to the one taken at grazing incidence. The shift observed is consistent with the existence of a gradual change in carbon coordination as one probes nearer the initial growth interface, and the observed binding energy approaches that of tetrahedrally coordinated carbon (287.28eV[24]). This effect is likely related to several features common to all plasma deposition methods. When one is depositing a highly insulating film upon a conductive substrate, the potential between the growth interface and the plasma is time varying as the insulating dielectric layer is deposited. Several workers have reported significant variations in resultant film structure with changes in bias[10], and this is likely the underlying cause of the anisotropy observed. To verify whether this inhomogeneity was accompanied by a variation in film stoichiometry, hydrogen profiling by nuclear reaction was employed.

The hydrogen content and its profile through the thickness of the DLC films was measured for PECVD DLC samples deposited at different temperatures using the $H(^{15}N\alpha,\gamma)C$ nuclear resonant reaction at 6.4 MeV. The hydrogen depth profiles were determined using nitrogen ions in the energy range 6.36 to 6.67 MeV. The results, of ±3% accuracy, are presented in Fig.3. As can be seen in the right side of the plots, there is a gradual increase of the hydrogen content at the beginning of the deposition of the film, i.e from the film/substrate interface. Bulk film hydrogen contents, 38% to 50%, depending on the deposition temperature, were only obtained after more than 400Å of film had been deposited. A similar behavior was reported previously by Angus et.al.[6] for films deposited by several methods. This behavior is likely also the result of a gradual decrease of the surface bias with increasing film thickness. An increase of the overall hydrogen concentration with decreasing bias has been reported by Weissmantel et.al.[26], and Moravec and Lee[27]. As shown in Fig.3 the bulk hydrogen content is about 38%, 42% and 50% for the DLC films deposited at 250 C, 150 C and 30C respectively, indicating a decrease in the hydrogen content in the film with increasing substrate temperature. It has been found that the films deposited at lower substrate temperatures contain a polymeric phase, estimated at about 8% of the carbon content, while no polymeric component was identified in the film deposited at 250 C[15].

The spatially resolved measurements of hydrogen concentration as well $C_{sub}1s$ XPS data show clearly that the properties of these carbon layers are spatially varying over the first few hundreds of Å of the layers. XPS data imply that more "diamond-like" carbon bonding is obtained at the initial growth interface, while more graphitic carbon is found further into the bulk of the film. As the dielectric properties of the growth interface are time varying with deposition time, so too is the near surface bias potential, which can account for this phenomena. The effect observed here highlights the need to employ a variety of bulk as well as spatially resolved techniques to accurately characterize the local atomic coordination of carbon in DLC films.

4.OPTICAL PROPERTIES

Typically, DLC layers are seen to be weakly absorbing in the visible spectrum, tending towards transparent in the infrared[5]. Their transparency makes DLC films good candidates as protective optical coating. As in other properties, controlling the hydrogen content of these layers is critical in controlling their optical properties. As noted already, removal of the hydrogen from hydrogenated DLC by heating to 700 C causes the loss of IR transparency[22]. Hydrogen content in PECVD films is a strong function of deposition temperature, and thus too is the measured optical gap. As seen in Fig.4, the optical gap shrinks rapidly for PECVD films prepared above 250C[28], decreasing from 2.1 to 0.9eV, over the deposition temperature span of 25-375C. To put the scale of this change in perspective, the electical resistivity of these layers decreased by over ten orders of magnitude for samples prepared at 325 versus 25C. Similar behavior of E_{opt} was observed by Jansen[19] in DLC films deposited by sputtering with an argon-hydrogen ion beam. As a result of hydrogenation, E_{opt} increased from 0.7eV at 0% hydrogen to 1.2eV at 35 at% hydrogen in the film. There are a wide range of optical properties reported for "similarly" prepared films, again indicative of how system sensitive such materials are. The optical gap measured for DLC films spans the range from 0.38eV to 2.7 eV[9] [29-32]. E_{opt} =0.38 eV was measured for H/C=1 in ion beam prepared films[9], while E_{opt} =0.75 was measured for very low hydrogen content in glow discharge films[29]. A further complication is the precise manner in which E_{opt} itself has been defined, in that a sharp edge is not observed in absorption data, requiring the empirical definition of an optical gap using a Tauc plot.

Both the real part **n** and the imaginary part **k** of the index of refraction, and their spectroscopic variation have also be found to be dependent on the preparation conditions and on the hydrogen content of the DLC films. Moravec and Lee[27] obtained refractive indices from 1.7 to 2.25 for PECVD materials by changing the ratio RF power/pressure during deposition. **n** and **k** have been measured in films deposited from acetylene at 300C [33] using spectroscopic ellipsometry between 2700 and 6000 Å. It was found that **n** = 1.87 ± 0.05 for the whole spectral range. After annealing in Ar at 400 and 500 C , **n** decreased and became dependent on photon energy. The films were both more dispersive and absorbing after each subsequent anneal, indicating the layers were again transforming into more "graphitic" material. For the same films, E_{opt} decreased from the initial value of 1.3 eV to 1.13 eV after annealing at 400C, and then to 0.85 eV after annealing at 500C[33]. These results differ from those obtained by thermal cycling DLC films to 500C in hydrogen, when no significant changes in the optical properties were observed[34]. It was argued in the second case that hydrogen from the annealing atmosphere had stabilized the films, thus preserving the diamond-like structure. Khan et.al.[34] also found **n** and **k** to be similar for both RF and ion beam sputtered films, but **n** was considerably lower and **k** higher for DC sputtered films. Non-hydrogenated DLC films prepared by sputtering of graphite showed a decrease of the index of refraction with increasing photon energy in the visible[35], while the hydrogenated ones deposited from acetylene by PECVD show an opposite behavior[36].

Smith modeled the optical properties of hydrogenated DLC using an effective medium approximation, assuming the films to be composed of amorphous diamondlike carbon, graphitic and polymeric carbon, and void components[30]. He found that the as-deposited films contained amorphous, polymeric, diamond like and graphitic components, with the transparency of the film resulting from the dominance of the polymeric and diamond-like components. After annealing at temperatures above 450C the graphitic component grows, a void component appears, while the polymeric and diamond-like components decrease to zero, thus degrading film transparency.

<u>5.ADHESION</u>

In order to perform their protective role, the DLC films have to adhere very well to the substrates, the adhesive forces having to overcome the high internal stresses in the films, which will otherwise cause the films to delaminate from the substrates. Similar to other properties the adhesion of the DLC films is dependent on preparation method and obviously on the substrate on which it is deposited.

A useful technique for measuring the adhesion of coatings to substrates is the scratch test[37], which determines the critical (minimum) load which induces damage to the coating. However for a multilayered structure, especially one made of thin films, it is also important to determine the location of the induced damage. The adhesion of DLC films to different substrates, or the adhesion between different layers, can be determined by a fracture test, where metal studs bonded to the surface of a structure are pulled at progressively increasing load until fracture occurs. Several commercial versions of such apparatus are available, one being the Sebastian Adherence Tester[38]. The fracture load is registered by the apparatus and is assumed to be a measure of the adhesion strength. In practice, however, imperfect bonding of the stud to the film surface can occurr, such that the load is applied to a significantly smaller area than the actual stud head surface. Thus, although quantitative data regarding fracture stress is questionable in these instances, a good qualititive comparison of the relative bond strengths in multilayered structures may be obtained.

Good adhesion of DLC carbon layers has been found on silicon, quartz, and carbide forming substrates such as iron alloys and titanium[7 12 39]. Employing DLC films, numerous metals which are prone to corrosion or abrasion may be protected. However, not all metals form stable adherent interfaces with DLC films, and the achievement of good adhesion can thus be the gating factor in the utility of DLC in this application.

Fig.5 presents a diagram of a layered substrate structure coated with DLC, the numbers indicating the different possible fracture locations. Results of the adhesion measurements performed with the Sebastian Adhesion tester on PECVD DLC films deposited from acetylene are presented schematically in Fig.6, in which the different layer configurations are indicated on the X-axis, while the Y-axis presents the location of the fracture surface observed for the specific configuration[40]. The numbers in the brackets indicate the deposition temperature in degrees centigrade. The 0 value on the Y-axis indicates that the DLC film had no adhesion at all to the underlying surface and it peeled off upon removal from the deposition apparatus.

Fig.6A & B indicate that the DLC films adhere well to the surface of Si, independent of substrate temperature during deposition. The adhesive force between the DLC and Si is higher than the cohesion of Si, and the fracture occurs in the Si substrate. Fig.6C shows that the DLC films had no adhesion at all when deposited upon the metallic layer, independent of the deposition temperature. The films spontaneously decomposed to powder, and could be removed from the substrate with a soft cloth.

It was shown previously[41] that amorphous silicon, a-Si, crystallizes in contact with certain metals at relatively low temperatures, forming silicides. Under thermal (non-plasma) conditions chromium silicide forms at 500C, while platinum silicide forms at 280C[41]. In that thermal spikes are produced by the ions bombarding the surface, it is reasonable to assert that thin interfacial layers of the same silicides will form in PECVD at lower bulk temperatures of the substrate. In this case, it may be expected that the a-Si film will form an intermediate bonding layer between the metallic surface and the hard DLC film. Better adhesion will be expected for the a-Si deposited at 250C than at lower substrate temperature, since the higher temperature will promote both interdiffusion and silicide formation.

It was found, as shown in Fig.7C that depositing an interfacial layer of a-Si (20-40Å thick) between the metal surface and the DLC film, in a structure such as shown in Fig.5, the adhesion of the carbon layer to the metal was improved significantly, and fracture occurred during the adhesion test either in the Si substrate, or between the metallic layer and Si substrate[40]. The carbon/a-Si/Me interface was the only film junction which remained intact during the tests, indicative of substantially superior bonding at that junction.

When thicker (>100Å) intermediate a-Si layers were applied, the adhesion test fracture occurred in the metallic film, or between the metallic film and the Si substrate, as shown in Fig.7D, indicating that the thicker Si layer on the metallic film actually degraded the adhesion of the metallic film to the substrate material[40]. It is known[42] that silicide formation is generally associated with large volume changes. When the thickness of the silicide layer is sufficiently large, these volume changes can result in large internal stresses, which will cause the degradation of the strength of the metallic layer[42] and causing the fracture in the adhesion test to occur in the metallic layer containing the silicides.

6.SUMMARY

Diamondlike carbon films can be prepared by a variety of deposition conditions and precursor materials. The films are characterized by a mixture of sp^3 and sp^2 carbon bond in an amorphous matrix and by a significant amount of hydrogen incorporated into them.
The properties of the DLC films are strongly dependent on the preparation method and conditions and, for a given preparation method, on the amount of hydrogen incorporated in the film. It is thus not possible to extrapolate the DLC properties from one system to another and one has to specifically characterize the DLC films for each deposition system.
DLC films can also have properties which are anisotropic near the interface with the substrate onto which they are deposited. This nonuniformity has to be taken into account when characterizing and attempting to employ such thin layers.
It has been shown that DLC layers deposited by PECVD have excellent adhesion on Si surfaces. In instances where poor adhesion is obtained on metallic surface, an interfacial adhesion layer of amorphous silicon can be employed to facilitate adhesion, specifically in those cases where the metal has a stable silicide phase.

----

*) Also at Materials Eng.Dept., Ben-Gurion University, Beer-Sheva,Israel, Incumbent of Eric Samson Chair in Steel Processing.

7.REFERENCES

1. S.Aisenberg and R.Chabot, J.Appl.Phys. **42** 2953 (1971)

2. L-P. Andersson, Thin Solid Films **86** 193 (1981).

3. A.Bubenzer, B.Dischler, G.Brandt, and P.Koidl, J.Appl.Phys. **54** 4590 (1983).

4. J.Zelez, J.Vac.Sci.Technol. **A1** 305 (1983).

5. N.Savvides and B.Window, J.Vac.Sci.Technol. **A3** 2386 (1985).

6. J.C.Angus, J.E.Stultz, P.J.Schiller, J.R.MacDonald, M.J.Mirtich and S.Domitz, Thin Solid Films **118** 311 (1984).

7. C.Weissmantel, K.Bewilogua,K.Breuer , D.Dietrich, U.Ebersbach, H.J.Erler, B.Rau and G.Reisse, Thin Solid Films **96** 31 (1982).

8. P.V.Koeppe, V.J.Kapoor, M.J.Mirtich, B.A.Banks and D.A.Gulino, J.Vac.Sci.Technol. **A3** 2327 (1985).

9. M.J.Mirtich, D.M.Swec and J.C.Angus, Thin Solid Films **131** 245 (1985).

10. C.Weissmantel, in Thin Films From Free Atoms and Molecules, ed. K.J.Klabunde, Academic Press, New York 153 (1985).

11. Hsia-chu Tsai and D.B.Bogy, J.Vac.Sci.Technol. A5 3287 (1987).

12. J.C.Angus, P.Koidl and S.Domitz, in Plasma Deposited Thin Films, eds J.Mort and F.Jansen, CRC Press Inc., Boca Raton, FL. U.S.A. 89 (1986).

13. K.Enke, H.Dimigen and H.Hubsch, Appl.Phys.Lett 36 291 (1980).

14. L.Holland and S.M.Osja, Thin Solid Films 58 107 (1979).

15. A.Grill,B.Meyerson, V.Patel, J.A.Reimer and M.A.Petich, J.Appl.Phys. 61(8) 2874 (1987).

16. E. R. Andrews, Prog. in Nucl. Magn. Spec. 8 1 (1971).

17. L. R. Sarles and R. M. Cotts, Phys. Rev. 111 853 (1958).

18. R.H.Jarman and G.J.Ray, J.Chem.Soc.Chem.Comun. 107 1153 (1985).

19. F.Jansen, M.Machonkin, S.Kaplan and S.Hank, J.Vac.Sci.Technol. A3 605 (1985).

20. S.Kaplan, F.Jansen and M.Machonkin, Appl. Phys. Lett. 47 750 (1985).

21. B.Dischler, A.Bubenzer and P.Koidl, Solid State Commun. 48 105 (1983).

22. M.P.Nadler, T.M.Donovan and A.K.Green, Thin Solid Films 116 241 (1984).

23. O. Matsumoto, H. Toshima, and Y. Kanzaki, Thin Solid Films 128 341 (1985).

24. T. Mori and Y. Namba, J. Appl. Phys. 55 3276 (1984).

25. K. Miyoshi and D. H. Buckley, Appl. of Surface Science 10 357 (1982).

26. C.Weissmantel, K.Breuer and B.Winde, Thin Solid Films 100 383 (1983).

27. T.J.Moravec and J.C.Lee, J. Vac. Sci. Techn. 20 338 (1982).

28. B.Meyerson and F.Smith, J.Non Cryst.Solids 35/36 435 (1980).

29. K.Pirker, R.Schallauer, W.Fallman, F.Olcaytug, G.Urban, A.Jachimowicz, F.Kohl and O.Prohaska,Thin Solid Films 138 121 (1986).

30. F.W.Smith, J.Appl.Phys. 55 764 (1984).

31. V.Natarajan, J.D.Lamb, J.A.Woolam, D.C.Liu and D.A.Gulina, J.Vac.Sci.Technol. A3 681 (1985).

32. S.Craig and G.L.Harding, Thin Solid Films 97 345 (1982).

33. Shuhan Lin and Shuguang Chen, J.Mater.Res. 2(5) 645 (1987).

34. A.Azim Khan, D.Mathine, J.A.Woollam and Y.Chung, Phys.Rev.B 28 7229 (1983).

35. S.F.Pellicori, C.M.Peterson and T.P.Henson, J.Vac.Sci.Technol. A4 2350 (1986).

36. D.R.McKenzie, R.C.McPhedran, N.Savvides and L.C.Botten, Phil.Mag. B48 341 (1983).

37. H.E.Hinterman, Wear 100 381 (1984).

38. QUAD Group, Santa Barbara, California.

39. C.Weissmantel, C.Schurer, F.Frohlich, P.Grau and H.Lehmann, Thin Solid Films 61 L5 (1979).

40. A.Grill, B.Meyerson, V.Patel, J.Mater.Res. 3 214 (1988).

41. S.R.Herd, P.Chaudhari and M.H.Brodsky, J.Non Cryst.Solids, 7 309 (1972).

42. S.P.Murarka, J.Vac.Sci.Techn., 17 775 (1980).

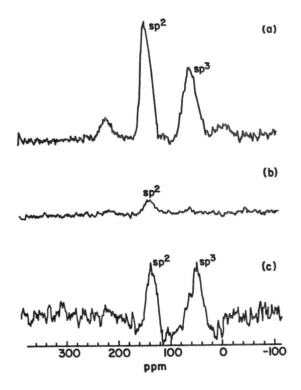

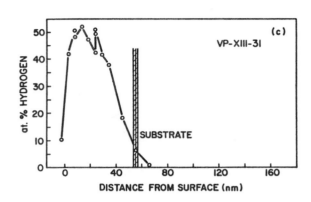

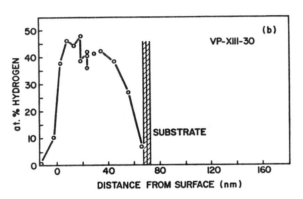

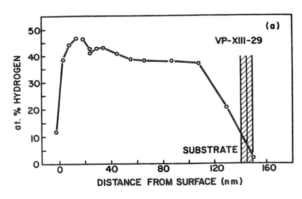

Figure 1. ^{13}C NMR spectra of DLC film deposited at 250C [15]. a) with proton decoupling; b) without proton decoupling; c) difference between spectra (a) and (b).

Figure 3.Hydrogen content profiles of PECVD DLC films deposited at different temperatures[15]. a) 250 C; b) 150C; c)30C.

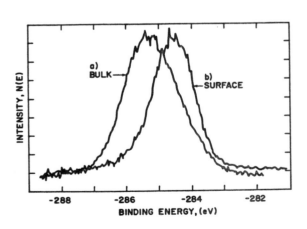

Figure 2. XPS carbon peak, C_{1s} from 40Å DLC film[15]. a) at 5° from surface's normal; b) at 75° from surface's normal.

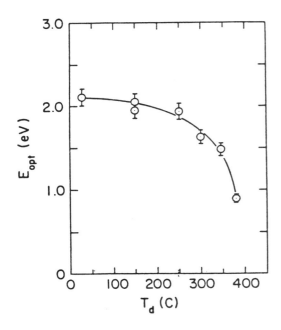

Figure 4.Optical gap of PECVD DLC films vs deposition temperature [28].

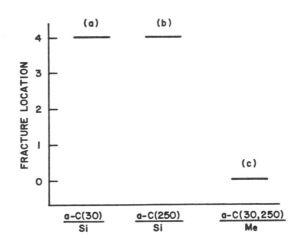

Figure 6.Adhesion fracture surfaces in layered structures with DLC[40]. The numbers on the y-axis correspond to surfaces indicated in Fig.5; the number in the brackets indicate the deposition temperature in degrees Centigrade.

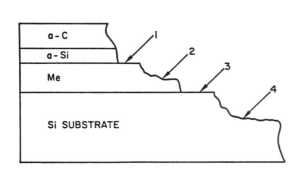

Figure 5.Diagram of layered structure; the numbers indicate possible fracture surfaces in adhesion test[40].

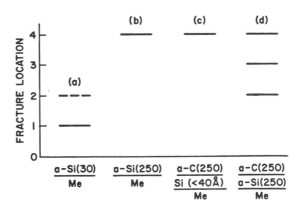

Figure 7.Adhesion fracture surfaces in layered structures with DLC[40]. See explanation in Fig.6.

Diamondlike carbon for infrared optics

Bhola N. De, S. Orzeszko[*] and John A. Woollam
The University of Nebraska-Lincoln, Department of Electrical Engineering
209N Walter Scott Engineering Center, Lincoln, Nebraska 68588-0511   U.S.A.

and

David C. Ingram[+] and A. J. McCormick
Universal Energy Systems
Dayton, Ohio 45432   U.S.A.

[*]On leave from Nicholas Copernicus University, Torun, Poland

[+]Now at Whickham Ion Beam Systems Ltd., Newcastle-upon-Tyne, Great Britain

1. INTRODUCTION

Diamondlike Carbon (DLC), also known as amorphous hydrogenated carbon (or a:C-H), is a hard semitransparent material. Since it is amorphous and has no grain boundaries, it is proposed for use as a coating on infrared optical surfaces to protect them from the environmentally adverse chemicals such as the acids, salt water, etc., as well as particle impact. Moisture penetration studies of DLC, using the Variable Angle Spectroscopic Ellipsometry (VASE), have been presented elsewhere [1,2]. In this paper, we first briefly mention sample preparation, done using various combinations of RF power and gas pressure in a plasma deposition chamber. Then, results of using an ultraviolet-visible (UV-VIS) absorbance spectrometer are presented. From the absorbance we calculated the band gap using the Tauc plots, and found values ranging from 0.2 to 1.25eV, depending on deposition condition.

Secondly, a series of DLC samples were prepared using a Kaufman type ion beam deposition system. Using VASE, we have studied these samples after bombarding them with various fluences of Fl, and analyzed the data using two Lorentz oscillators to represent the optical properties. We have also shown the correlation between the optical properties and the amount of hydrogen present within the samples, as measured by proton recoil experiments.

2. EXPERIMENTAL

Samples were deposited on silicon, quartz, KG-3 glass, BK-7 glass, heavy metal fluoride (HMF) glass, ZnS, and lexan. Before depositing on these samples we went through an extensive substrate cleaning procedure as follows: First, the substrate was ultrasonically cleaned in 1,1,1-Trichloroethane. It was then cleaned with acetone. Next, it was washed with methanol, and then washed with deionized water, and finally it was dried by blowing with dry nitrogen. For lexan, the first two steps were omitted and for the HMF glass, only the last step was used, because of the adverse chemical reaction of these substrates with those particular solvents.

The schematic diagrams of two different deposition systems that were used are shown in Fig. (1). In the first case, a 13.6 MHz RF power source was used to excite the gas plasma. A 1:1 mixture of methane and argon was used for the gas source. Since the plate areas were different, there was a substantial bias voltage that resulted in the sputtering of the substrate and the sample; this caused some heating of the substrate. The possible consequence of this effect is the change in the bandgap energy of DLC, as will be described later. We found that the film spalled off the ZnS substrate if the film thickness reached more than about 100Å. The DLC depositions on Si substrates were the most successful. We used DLC deposited on quartz substrates for the UV-VIS absorbance study, and from Tauc plots of these data the optical bandgap energy was extracted.

In the second preparation method, a Kaufman type ion beam deposition system was used to deposit DLC on silicon substrates. We used methane gas for the carbon source, and the ion beam energy was about 500eV. The samples were then irradiated with the following fluences of Fl (at 6.4 MeV): 3×10^{14}, 1×10^{15}, 3×10^{15} and $1 \times 10^{16} cm^{-2}$. These samples were then used for the VASE study.

3. STUDY OF SAMPLES PREPARED BY RF PLASMA

Fig. (2) shows the optical gap of DLC vs. power, for different background gas pressures. Notice a steady decrease of the bandgap energy as the power is increased, especially for higher gas flow rates. The bandgap energy spans the range from 0.25eV to 1.25eV. Since the dc self-bias voltage at higher power level is higher, the substrate heating at the surface is higher due to the higher impact energy of ions onto the substrate. This result is consistent with published results[3], according to which as the substrate temperature increased, the bandgap energy correspondingly decreased.

4. STUDY OF SAMPLES PREPARED BY ION BEAM

We have used VASE to study the optical properties of as-deposited and Fl beam modified DLC as a function of fluences of Fl. For details on VASE, the reader is referred to the literature[4-10]. Briefly, a linearly polarized beam of light interacts with the sample and becomes elliptically polarized. The ellipticity depends on the optical properties and thicknesses of different layers interacting with the light. By measuring the ellipticity, we obtain the ellipsometric ψ and Δ parameters, defined by the ratio of the complex s-wave and p-wave Fresnel reflection coefficients:

$$p = \frac{R_p}{R_s} = \tan \psi \exp j\Delta. \tag{1}$$

By modeling the interaction of light with these different layers and fitting the theoretical ψ, Δ values with the experimental data we obtain the optical constants and the thicknesses of different layers. In fitting the experimental data, one looks for the lowest mean-square error (MSE), defined by:

$$MSE = \frac{1}{N} \sum_{i=1}^{N} \left((\psi_i^{exp} - \psi_i^{cal})^2 - (\Delta_i^{exp} - \Delta_i^{cal})^2 \right). \tag{2}$$

The optical constants of our DLC samples were represented by two Lorentz oscillators. It is well known that[11] the dielectric constant for M Lorentz oscillators is given by:

$$\epsilon = 1 + \sum_{i=1}^{M} A_i \left(\frac{1}{E + P_i + jW_i} - \frac{1}{E - P_i + jW_i} \right), \tag{3}$$

where the constants A_i, P_i, and W_i are the amplitude, center frequency, and width respectively for the i-th oscillator and E is the photon energy. By using Lorentz oscillators, the number of unknown parameters is greatly reduced in the data analysis as long as the total number of Lorentz oscillators is kept small. Also, this Lorentz oscillator model presents us with a basis for predicting the optical constants of DLC outside the measured wavelength range.

The ellipsometric parameters were measured at three different angles: 60°, 65° and 70° and over the wavelength range from 3000Å to 8000Å, at 100Å intervals. The ellipsometric data were then analyzed using a model with substrate and single layer film. Data analysis gives the Lorentz oscillator parameters as a function of fluence, as shown in Figs. (3) and (4). Fig. (3) shows the oscillator position, and Fig. (4) shows the oscillator amplitude. Except for the second oscillator amplitude, there is a definite trend in oscillator parameters as a function of the Fl dose. We propose that the lower energy oscillator represents the $\pi \rightarrow \pi*$ band transition, and the higher energy oscillator represents the $\sigma \rightarrow \sigma*$ band transition of DLC.

Proton recoil experiments were carried out on these samples to obtain the concentration of hydrogen as a function of the Fl dose and also to determine the relationship between hydrogen content and the energy bandgap of DLC. The approximate energy bandgap was calculated from the Tauc plot, derived from the calculated absorbance using the Lorentz oscillator model for DLC. (Strictly speaking, this procedure is not proper, but it does give a rough indication.) The result of the measured hydrogen concentration as a function of the Fl dose is shown in Fig (5); the initial rate of hydrogen loss is fast but it reaches a steady state value of about seven percent. In Fig. (6), is plotted the energy gap as a function of hydrogen content. Notice that the gap decreases as hydrogen is removed from the sample.

5. CONCLUSION

We have prepared DLC samples by two different methods: RF gas plasma, and Kaufman type ion beam deposition.

The first method was used for the study of energy bandgap of DLC as a function of different combinations of power and pressure. The measurements indicate a range of bandgaps of 0.25eV - 1.25eV. We attributed this effect to surface heating of the substrate caused by the sputtered ions.

The second method was used for studies of the relationship between the optical properties of DLC and the dosage of implanted Fl ions. We used VASE and a model of two Lorentz oscillators to study the optical properties as a function of the Fl dose. Using proton recoil, the content of hydrogen in DLC was determined for different fluences of Fl and it was found that increasing amounts of hydrogen were removed with increased Fl dosage. The initial removal rate was rapid, until the hydrogen concentration reached a value of about seven percent. Removal of hydrogen caused a decrease in bandgap energy.

6. ACKNOWLEDGEMENT

This research was supported by the U.S. Army Materials and Analysis Laboratory, Contract no. DAAL04-86-C-0030.

7. REFERENCES

1. S. Orzeszko, B. N. De, P. G. Snyder, J. A. Woollam, J. Pouch and S. A. Alterovitz, Journal de Chimie Physique **84**, 1469 (1987).
2. S. Orzeszko, B. N. De, J. A. Woollam, J. Pouch, S. A. Alterovitz and D. C. Ingram, J. Appl. Phys., in press (1988).
3. J. C. Angus, P. Koidl and S. Domitz, "Carbon Thin Films", in Plasma Deposited Thin Films, Chemical Rubber Co., Preso Publishers, Boca Raton, FL, 1986.
4. R. M. A. Azzam and N. M. Bashara, Ellipsometry and Polarized Light, North-Holland Publishing Co., New York, 1977.
5. D. E. Aspnes, "Characterization of Materials, Interfaces and Laminar Structures by Optical Spectroscopic Techniques" in SPIE Symposium on Microlithography, SPIE, Bellingham, WA, 1986.
6. D. E. Aspnes, J. Vac. Sci. Technol. **18**, 289 (1981).
7. K. Vedam, MRS Bulletin **12**, 21 (1987).
8. G. H. Bu-Abbud, N. M. Bashara and J. A. Woollam, Thin Solid Films **38**, 27 (1986).
9. P. G. Snyder, M. C. Rost, G. H. Bu-Abbud, J. A . Woollam and S. A. Alterovitz, J. Appl. Phys. **60**, 3293 (1986).
10. J. A. Woollam, P. G. Snyder, A. W. McCormick, A. K. Rai, D. Ingram and P. Pronko, J. Appl. Phys. **62**, 4867 (1987).
11. M. Erman, J. B. Theeten, N. Vodjdani and Y. Demay, J. Vac. Sci. Technol. B1, 328 (1983).

DLC Deposition Chambers

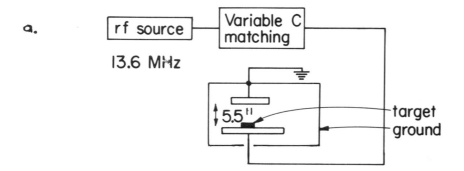

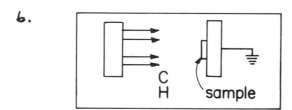

Figure 1. Schematic diagram of (a) 13.6 MHz RF power driven gas plasma, and (b) Kaufman type ion beam deposition system for DLC.

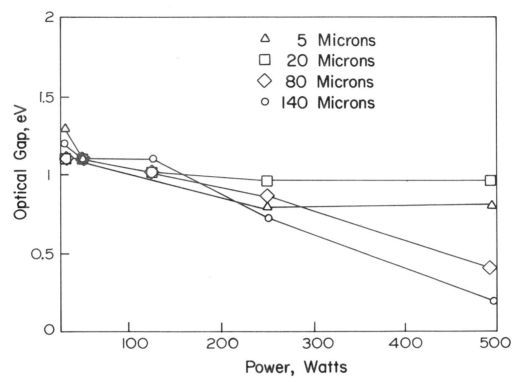

Figure 2. Plot of band gap energy of as-deposited DLC on quartz substrates vs. the RF power.

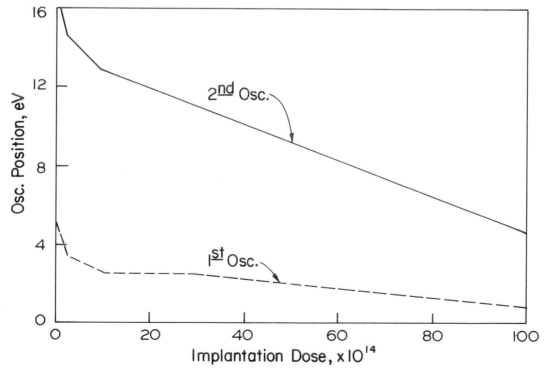

Figure 3.   Plot of the positions of Lorentz oscillators vs. fluence of F1.

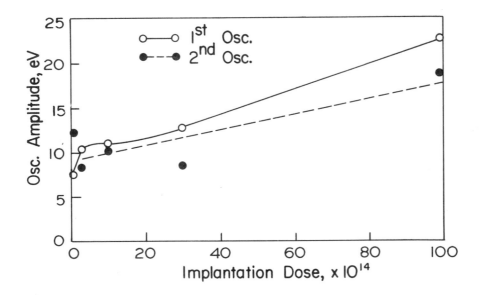

Figure 4. Plot of the amplitude of Lorentz oscillators vs. fluence of Fl.

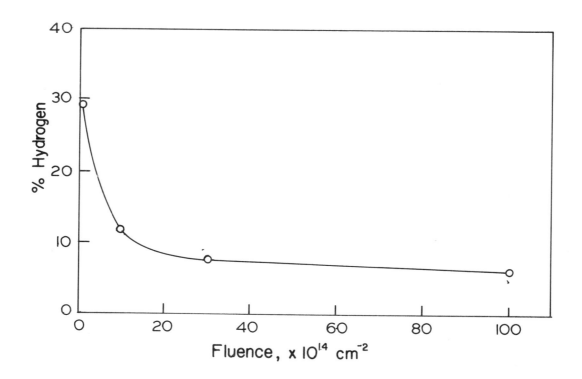

Figure 5. Plot of the relative hydrogen concentration in DLC vs. Fl dose, as determined from proton recoil experiment.

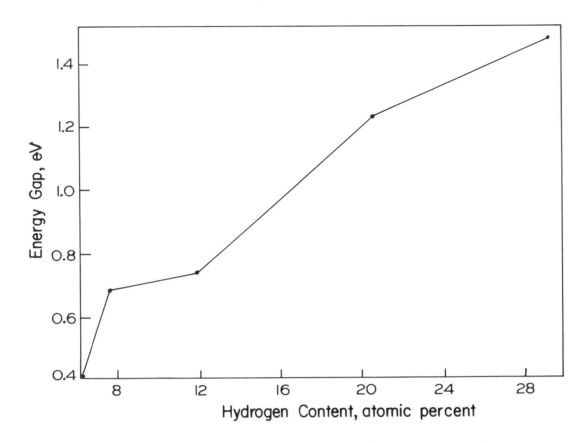

Figure 6. Plot of energy gap vs. hydrogen content.

Growth of diamond by rf plasma assisted chemical vapor deposition

Duane E. Meyer, Natale J. Ianno, and John A. Woollam

Department of Electrical Engineering, University of Nebraska
Lincoln, Nebraska 68588-0511

A. B. Swartzlander and A. J. Nelson

Solar Energy Research Institute
Golden, Colorado 80401

ABSTRACT

Diamond particles have been produced by inductively coupled radio frequency plasma assisted chemical vapor deposition. Analysis indicates that particles having a thin graphitic surface, as well as diamond particles with no surface coatings have been deposited. Scanning electron microscopy analysis shows that particles are deposited on a pedestal which Auger spectroscopy indicates to be graphitic. This phenomenon has not been previously reported in the literature.

1. INTRODUCTION

The production of diamond from low pressure, low temperature processes has become an important technology as evidenced by the rapidly growing number of papers being published on this subject.[1-9] Recently the largest research effort has been in the area of microwave plasma and filament assisted CVD processes, and results are generally characterized by Raman spectroscopy.[1-4] For rf plasma processes there is need for rf powers greater than 500 watts and substrate temperatures generally greater than 800°C.[5,6] In addition, electron assisted CVD, thermal plasma assisted CVD, and filament assisted microwave plasma CVD have also been investigated.[7-9] In this paper we report on the preparation of diamond from CH_4/H_2 gas mixtures via plasma assisted CVD, and the analysis of the resulting material.

2. EXPERIMENTAL

A schematic diagram of the apparatus is shown in Fig. 1. The rf generator inductively supplied up to 500 watts at 13.56 MHz. The reaction tube was quartz with a diameter of 1.4 cm and a length of approximately 30 cm surrounded by a concentric cooling tube of quartz with a diameter of 25 mm and a length of 20 cm in a horizontal geometry. The plasma was confined to a length of approximately 20 cm within the reaction tube using quartz wool plugs at either end of the reaction tube.

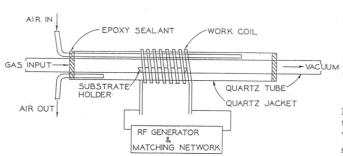

Figure 1: Quartz reaction chamber design.

Deposition pressures ranged from about 0.5 to 35 torr. Methane concentrations ranged from 0.25 to 2.0 percent in hydrogen, and total flow rates were between 8.2 and 100.0 sccm. Input powers ranged between 200 and 500 watts and substrate temperature ranged from less than 800°C to 1030°C. Deposition times were typically 2 hours but ranged from as short as 30 minutes to as long as 41 hours.

The substrates were typically 0.5 cm x 0.7 cm. When a graphite susceptor was used for heating, a conductive carbon paint was used to bond the substrate to the susceptor to facilitate a good thermal contact. The surfaces of silicon substrates were either polished or roughened by sandpaper, sandblasting, or Si-Si abrasion. No surface modification was attempted on the glass, quartz, nickel or boron nitride substrates. In all cases, the substrates were cleaned by sonicating in trichloroethylene and acetone. Notice that the system as presently configured was heated by the plasma, and this depended on substrate material and geometry.

3. RESULTS AND DISCUSSION

Deposition on glass, quartz, boron nitride and nickel were only partially successful in forming diamond particles. The use of silicon substrates resulted in the best particle coverage. Deposition of particles of various sizes and shapes were observed in all cases. The surface density of the deposited particles also varied from being very sparse (a few percent) to being relatively dense (perhaps 70 percent of surface coverage).

Polished surfaces resulted in the most sparsely populated particle deposits. In fact, deposits were virtually nonexistent except in areas of crystal defects such as cracks or edges where a roughened surface existed due to microcleavage of the silicon wafer. This was true except in the case where a layer of non-crystalline carbon was deposited on the silicon surface. In this case, particles formed on top of the carbon layer surface with a much more dense population.

Silicon surfaces which were sandblasted with aluminum oxide and abraded by rubbing two silicon wafers together also exhibited poor partical population densities. The particles deposited on sandblasted surfaces were roughly spherical in shape and typically 1.5 μm in diameter. Particles deposited on the abraded surface had a population density larger than that observed on the silicon surface types described above. The particles appeared to be somewhat egg-shaped and were typically about 1.5 μm in diameter.

Silicon surfaces which were sandpapered produced the largest particle population densities, for deposition parameters similar to those used for deposition on the silicon surface types described above. The largest number of depositions were made using this type of substrate, and a range of deposition parameters were investigated. When the HOPG holder was used, the substrates were cemented to the holder with carbon paint, and the substrate temperature varied from about 830°C to about 1000°C. Deposition parameters for sandpapered Si surfaces are given in Table 1. In all cases the overall general shape of the deposited particles was the same. The particle sizes ranged from about 0.5 μm to about 7.5 μm, with varying degrees of other carbon deposits observable. In all cases when substrate heating was used, analysis by Auger microprobe spectroscopy resulted in nearly identical spectra. The first derivative C-KLL spectrum of a typical particle is discussed below.

Table 1: Deposition parameters for samples on sandpapered Si, using HOPG holder

Sample	CH₄ %	Pressure torr	Flow Rate sccm	RF Power watts	Time hours	Temp. degC
35	2.0	10.0	100.0	400	2.8	930
43	0.5	24.6	49.8	400	3.0	995
44	0.5	24.6	50.0	500	3.8	980
47	0.5	10.2	50.0	350	2.2	950
48	0.5	26.6	50.0	500	2.0	970
50	0.5	34.6	50.0	500	2.0	990
51	1.0	10.3	25.0	325	3.5	830
52	1.0	10.0	25.0	500	5.0	1000
58	0.5	40.5	50.0	500	33.0	900
60	0.5	15.4	50.0	230	18.0	1000

Two depositions were made on sandpapered Si substrates without using a substrate holder. The substrate was placed on a quartz boat within the reaction tube at about the center of the induction coil. The substrate temperature was not measured in these depositions because they were colder than the lower limit of 700°C for the optical pyrometer. The deposition parameters for samples deposited using no substrate holder are given in Table 2.

Table 2: Deposition parameters for samples on sandpapered Si, using no substrate holder

Sample	CH₄ %	Pressure torr	Flow Rate sccm	RF Power watts	Time hours	Temp. degC
61	0.5	15	50.5	500	15.5	UNK
62	1.0	15	50.5	500	41.2	UNK

The SEM study shows that both egg-shaped and faceted particles were deposited. It should be noted that the faceted particles were deposited when there was no substrate heating, while only the egg-shaped particles were deposited when substrate temperatures were between 800°C and 1000°C. This is in contrast to reports that substrate temperatures between 800°C and 1000°C are optimal for diamond formation from plasma assisted deposition processes.[9] It can be observed from Table I that the egg-shaped particles were deposited when deposition temperatures were between 800°C and 1000°C for a range of other deposition parameters.

Additionally, it can be observed that a small pedestal forms beneath each of the egg-shaped particles (Fig. 2).

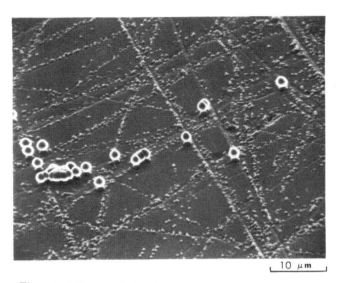

Figure 2: Micrograph showing pedestals under the particles.

Fig. 3a is a micrograph which shows a region where particles embedded in a layer of carbon deposits has flaked away and Fig. 3b is a blow-up of this region showing the pedestals which have been left behind. An Auger microprobe analysis of these pedestals was done and and it is interesting to note the absence of any detectable amount of silicon in the pedestal.

Fig. 4 compares the first derivative C-KLL spectrum of the pedestal with a spectrum of graphite.[10]

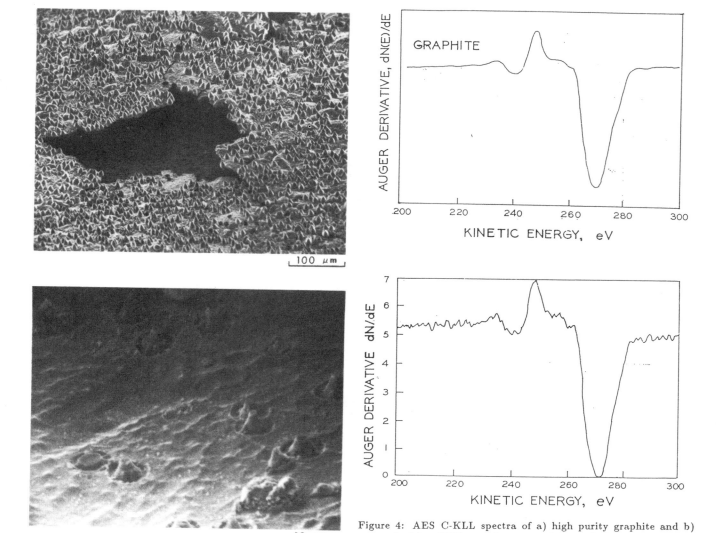

Figure 4: AES C-KLL spectra of a) high purity graphite and b) pedestal. After Williams, Moravec, and Orent [10].

Figure 3: SEM micrographs of (a) particles deposited in a layer of carbon deposits and (b) blow-up of central region on (a) showing the pedestals.

The Auger study indicates that the egg-shaped particles exhibit a C-KLL lineshape characteristic of a graphitic structure in all cases. This, however, is only an indication of the structure of the surface of the particles since the escape depth of a 270 eV (carbon KLL line) Auger electron is only about 7Å. Therefore, it is possible the egg-shape could result from the final deposition of graphitic carbon on top of a diamond particle, smoothing out any faceting which would exist. The faceted particles exhibit C-KLL lineshapes characteristic of diamond.[10,11]

4. CONCLUDING REMARKS

The analysis of the egg-shaped particles indicates that they at least have a graphitic surface layer. Additionally, it has been determined from our SEM studies that these particles always develop on a pedestal region. This is a phenomenon which has not been previously reported. Subsequent Auger analysis indicates that the pedestals have a graphitic structure. The fact that these particles only deposit on a graphitic pedestal may provide some clues to the kinetic processes going on during their formation. This may also shed some light on the process of diamond nucleation.

Auger microprobe analysis of the faceted particles indicates that these have a diamond structure. These particles were only deposited when no substrate heater was used and only the egg-shaped particles were deposited when a substrate heater was used. This indicates that the optimal depositon temperature may be somewhat dependent upon the configuration of the deposition system for an rf plasma assisted process.

5. ACKNOWLEDGMENTS

Duane Meyer was supported by a teaching assistantship from the University of Nebraska.

6. REFERENCES

1. S. Matsumoto, Y. Sato, M. Tsutsumi and N. Setaka, J. Mater. Sci., **17**, 3106 (1982).

2. Y. Sato, S. Matsuda and S. Nogita, J. Mater. Sci. Lett. **5**, 565 (1986).

3. A. Badzian, B. Simonton, T. Badzian, R. Messier, K. E. Spear and R. Roy, Proc. SPIE, **683**, 127 (1986).

4. N. Setaka, *Tenth International Conference on Chemical Vapor Deposition*, edited by G. W. Cullen (The Electrochemical Society, Inc., Pennington, NJ, 1987) pp. 1156.

5. S. Matsumoto, J. Mater. Sci. Lett., **4**, 600 (1985).

6. K. Kijima, S. Matsumoto and N. Setaka, Proc. Int'l. Ion Engineering Congress, **3**, 1417 (1983).

7. A. Sawabe and T. Inuzuka, Appl. Phys. Lett., **46**, 146 (1985).

8. S. Matsumoto, M. Hino and T. Kobayashi, Appl. Phys. Lett., **51**, 737 (1987).

9. A. R. Badzian, T. Badzian, R. Roy, R. Messier and K. E. Spear, Mat. Res. Bull., **23** (3), 385 (1988).

10. T. J. Moravec and T. W. Orent, J. Vac. Sci. Technol., **18** (2), 226 (1981).

11. A. Koma and K. Miki, Appl. Phys. A **34**, 35 (1984).

SESSION 4

Characterization II

Chair
W.-K. Chu
University of North Carolina

Growth of diamond-like films on Ni
surfaces using remote plasma enhanced chemical
vapor deposition

R.A. Rudder, J.B. Posthill, G.C. Hudson, M.J. Mantini, and R.J. Markunas

Research Triangle Institute, Semiconductor Research

P.O. Box 12194, Research Triangle Park, North Carolina 27709

ABSTRACT

Remote plasma enhanced chemical vapor deposition has been used to investigate the nucleation and growth of diamond and diamond-like films on Ni surfaces. The primary results center around hydrogen dilution experiments. Hydrogen dilution when using the polycrystalline Ni substrates tends to reduce the growth rate and increase the electrical resistivity of the films ($^{-}10^7$ Ω-cm), it is found that at even higher hydrogen dilutions (greater than 98% H_2) the films become semicontinuous with sparse and sometimes no nucleation occurring. These films, like the ones grown at lower hydrogen dilution, do not show a 1332 cm^{-1} diamond Raman line but show graphitic and disordered carbon features. An attempt to grow heteroepitaxial diamond on Ni(111) surfaces under conditions of high hydrogen dilution (100:1) produced a sample with oriented hillocks which are heteroepitaxially in registration with the substrate. Raman analysis showed lines characteristic of graphite and disordered carbon with an additional line at 1398 cm^{-1}. Transmission electron microscopy produced a diffraction pattern with the lattice spacing and symmetry of epitaxial graphite with some faint polycrystalline rings.

1. INTRODUCTION

Diamond is an exciting electronic material. The high electron and hole mobility values observed in natural diamond combined with its high thermal conductivity make it an excellent candidate for high power, high frequency devices. However, before its properties can be practically exploited, a semiconducting heteroepitaxial diamond technology must be developed. To date, no heteroepitaxial diamond growth has been reported. The majority of work reported is for the growth of diamond polyhedra on scratched silicon and refractory metal surfaces.[1-3] There, the work has not focused on epitaxial growth of diamond for semiconducting properties, but rather has focused on growth of semicontinuous polycrystalline films. In coatings applications, scratching the surfaces prior to diamond growth would appear to be a successful approach for increasing the nucleation density and producing a more uniform film. It seems unlikely, however, that a heteroepitaxial technology could be realized using such an approach.

Our approach to the problem has been to develop an heteroepitaxial technology based on a lattice matched $Ni_{1-x}Cu_x$ substrate system. An ultra-high vacuum transfer system couples a surface analytical unit, a metals MBE unit, a substrate preparation chamber, and a remote plasma enhanced chemical vapor deposition (RPECVD) system. A prerequisite for any epitaxial process is the presentation of a clean, ordered surface to the deposition process. In this approach the MBE, substrate preparation, and surface analytical unit are to determine the surface structure and chemistry prior to diamond nucleation, and growth. Besides dealing with substrate issues which can greatly influence the nucleation process, in diamond epitaxy, one must also be concerned with minimizing the number of incorporated sp^2 bonds. The RPECVD system was the system first chosen to address the nucleation issue[4]. This technique was chosen not because it was a proven technique for growing diamond, but rather it was a technique for selectively creating various hydrocarbon radicals. This was based on earlier work by Balumata etal.[5] and Bolden etal.[6] that showed different CH_4 branching patterns for collisions with He, Ar, and Xe metastables. The branching patterns range from chemionization pathways with He metastables to simple fragmentation into methyls with Xe metastables. The advantages of this overall approach are (1) more clearly defined growth surfaces upon which nucleation can be studied and (2) more clearly defined fluxes of reactive species impinging on the surface during the nucleation and growth process.

This article reports on RPECVD growths on polycrystalline Ni substrates and a growth on single crystal Ni(111). It critically assesses both the quality of the material produced and the implementation of the approach. This work was performed prior to completion of the ultra-high vacuum transfer system. All the substrates used in this study had an in-situ hydrogen treatment for a duration of 10 min. The intent was to reduce the nickel oxides prior to carbon deposition. There was no in-vacuo capability to determine the effectiveness of this procedure for the experiments reported here. Subsequent work has shown the hydrogen treatment capable of removing oxides from Ni surfaces. The primary experimental variable reported here is hydrogen dilution of the methane. Optical emission results demonstrate that both atomic hydrogen and CH_x radicals can be produced downstream from a noble gas plasma discharge. Remote activation of the hydrogen was necessary to avoid silicon incorporation into the film from plasma tube wall erosion. The results observed for higher hydrogen dilutions using the polycrystalline Ni substrates were a reduction in growth rate, an increase in electrical resistance, and a reduction in the π-bonding. No 1332 cm^{-1} diamond Raman line was observed on any of the samples. Growth on a single crystal Ni(111) surface produced a carbon layer whose surface symmetry showed registration between the carbon layer and the Ni(111) substrate. However, Raman scattering and transmission electron microscopy (TEM) data reveal the material to be predominantly epitaxial graphite.

2. EXPERIMENTAL RESULTS AND DISCUSSION

2.1 Remote plasma enhanced chemical vapor deposition:

As part of our approach, RPECVD was chosen as the tool for investigating the growth and nucleation of diamond[4]. Figure 1 shows a schematic of the RPECVD reactor. There are two primary differences between RPECVD and conventional PECVD. First, the reagent gas molecules are not excited in the plasma region but instead react with excited gas species or electrons flowing from the plasma region. The plug velocity of gas through the plasma tube and the separation distance from the plasma coil to the ring feed prohibit back diffusion of the reagent into the plasma discharge. The metastable species pumped from the discharge are 4 - 20 eV above their ground state depending on the metastable gas atom selected. Upon their interaction with a reagent gas, a fairly limited number of fragmentation or ionization products are produced. Second, unlike conventional PECVD, the substrates are well removed from the plasma region, minimizing the plasma densities near the substrate. This should result in small sheath fields between the substrate and the plasma (in contrast to immersion systems). Ions created by Penning processes in the vicinity of the substrate see no large electric fields to accelerate them and will undergo collisions with other molecules before arriving at the substrate.

2.2 Chemiluminescence studies to qualify metastable flux:

In order to take advantage of the chemical selectivity of the various noble gas metastables, one must develop a "clean" source of metastables. Typically, metastables are produced in fairly low power discharges. Higher power discharges result in rapid depopulation of the metastable flux due to promotion to higher electronic or ionic states.[7]. If a metastable source contains large fluxes of ions or electrons, those components result in different chemical branching producing unintentional fluxes of other radicals to the growth surface. Therefore, a means of verifying metastable flux into the reaction zone is necessary. One reported way to detect the presence of metastables is through the use of a chemiluminescent reagent such as N_2.[7-8] The interaction of N_2 with Ar metastables (Ar[*]) produces an excited electronic nitrogen molecular state ($C^3\Pi_u$) which relaxes to a lower electronic molecular state ($B^3\Pi_g$) with the emission of photons characteristic of the transitions between the various vibro-electronic states. We have exploited such reactions to verify the production and transport of metastables from the plasma region downstream to the reaction zone.

Shown in Figure 2 is an emission spectrum taken from the ring area downstream from an Ar discharge with N_2 flowing from the ring feed. The plasma conditions are such that the plasma is confined inside the rf coils. Visibly, the emission has a characteristic violet color which has been observed by others using dc plasma discharges. The spectrum delineates transitions of the nitrogen second positive system Other workers have obtained similar emission spectrum from N_2 introduced in an Ar afterglow.[7,8] Furthermore, these authors established that these lines were characteristic of interactions between Ar[*] and N_2. This spectrum is evidence that metastables created in the discharge region are transported down into the reaction zone. The numbers in

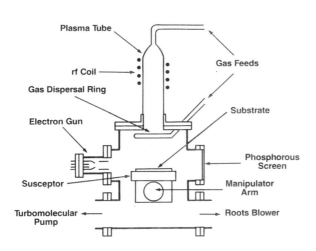

Figure 1.   Schematic of Reactor

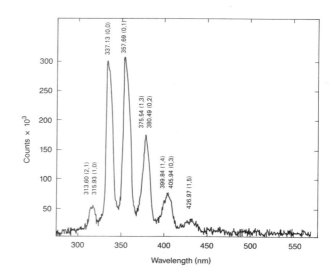

Figure 2.   Metastable Ar excited
                 N$_2$ emission spectrum and
                 hot filament gas doser

parenthesis refer to transitions from the upper C $^3\Pi_u$ vibrational levels to lower B $^3\Pi_g$ vibrational levels, respectively. Note that the Ar* levels (11.55 and 11.74 eV), fall between the second (11.51 eV) and the third (11.75 eV) vibrational levels of the C $^3\Pi_u$ electronic level.

Diamond deposition attempts using Ar with a confined plasma configuration resulted in minimal deposition. The deposition conditions were determined by varying reactor power, pressure, and flow rate conditions to maximize the N$_2$ chemiluminescence. After that determination, a sample was moved on to the heater stage, and CH$_4$ was introduced in the place of the N$_2$. No CH emissions were seen in the reactor. This would be consistent with the work of Balumata[5] that showed the branching for Ar* to be CH$_3$ and CH$_2$. Those species can not be detected by emission. The reactor flow conditions were 0.200 Torr, 500 sccm Ar (plasma), 500 sccm N$_2$(ring), and 10 W rf power. X-ray photoelectron spectroscopy revealed that the surface had probably less than 10 Å of carbon on it. Given that typical metastable concentrations range from the $10^{10} - 10^{11}$ cm^3 concentrations and given the excess CH$_4$ concentration to consume that flux, then it seems plausible that the slow growth rate may be a consequence of a rate limiting step such as H desorption from a hydrogen terminated surface. However, without quantification of the metastable flux and knowledge of methyl and methylene sticking coefficients, it is impossible to be certain.

Having been unsuccessful in depositing material under a confined discharge with only metastables entering the reaction zone, discharge conditions were changed such that the plasma region extended downward from the coils to a region slightly above the ring feed. Since transport of metastables is restricted by an exponential decay in time, having the plasma extend toward the reaction zone should shorten the pumping distance and thus reduce the time and increase the flux of metastables into the reaction zone. Perhaps, this would even allow some metastable flux at the substrate permitting dehydrogenation of the carbon surface. Figure 3 shows an emission spectrum from N$_2$ being excited under these conditions. Visibly, the emission has a characteristic orange color. Spectrally, one still observes many but not all of the transitions around 350 nm that have heretofore been attributed to Ar*. Besides these transitions, there are also transitions around 580 and 650 nm (not shown) of higher intensity than the second positive transitions. The additional transitions are from the first positive system of molecular nitrogen (B $^3\Pi_g$ - A $^3\Sigma_u^+$). Since they originate on a level lower in energy than the Ar* levels, and since their intensities are larger than the second positive transitions, there would seem to be a second energy mechanism for populating the B $^3\Pi_g$ state. Electron flux into the reaction zone is the most probable candidate. Other candidates, such as Ar$^+$ would have energies greater than the metastable energy and therefore be less likely to populate the B $^3\Pi_g$ state. Under the extended plasma conditions, replacing the N$_2$ with CH$_4$ resulted in rapid growth of carbon (~2000 Å/hr). Shown in figure 4 is the emission spectrum taken from the ring area showing the 4300 Å CH system. The layers, however, were opaque and electrically very conductive.

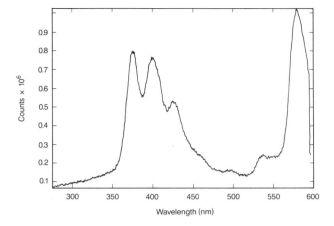

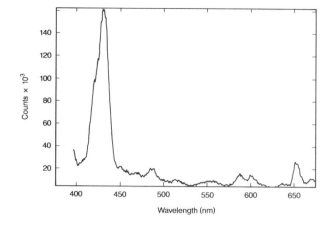

Figure 3.   Emission spectrum of
             N_2 excited by unconfined
             Ar plasma

Figure 4.   Emission spectrum of
             CH excited by unconfined
             Ar plasma

2.3 Deposition under hydrogen dilution conditions:

A series of experiments were then directed at adding hydrogen into the reactor. The first of these experiments produced carbon layers containing as much as 25% Si. At this time hydrogen was being introduced into the reactor through the plasma tube. Chemical erosion of the borosilicate glass walls in the immediate vicinity of the coils was responsible for the Si contamination. To circumvent this problem, the hydrogen was removed from the plasma gas feed and introduced downstream through the ring feed. The use of an unconfined Ar discharge produces some atomic hydrogen, but the dominant emission feature is a "blue" continuum due to transitions from an antibonding H_2 state. Balumata etal.[5] have reported that about 5% of the antibonding H_2 dissociates into atomic hydrogen. Recently, we have found that the use of an unconfined He discharge produces more atomic hydrogen relative to an Ar discharge. Currently, no quantification of the atomic hydrogen concentration has been made.

2.4 Growth on Ni substrates.

Polycrystalline Ni gaskets were used for deposition under a variety of hydrogen dilutions for Raman analysis. Depositions occurred using Ar as the plasma gas with various mixtures of CH_4 and H_2 flowing from the ring feed. The pressure in the reactor was maintained at 0.200 Torr, the substrate temperature was maintained at 650 °C, the Ar flow rate was maintained at 250 sccm, and the rf power at 50 W. The primary effect noticed with the addition of hydrogen was a reduction in growth rate from ⁻20,000 Å/hr with no hydrogen to ⁻200 Å/hr with 98% H_2. The growth rate could be increased by increasing the rf power delivered to the Ar discharge.

Unlike immersion discharge systems or hot filament reactors, increasing the hydrogen dilution did not result in a transformation of the films from disordered carbon and graphite into diamond. Raman analysis shown in figure 5 shows the primary effect of the hydrogen dilution is to minimize the disordered carbon peak with no appearance of the 1332 cm⁻¹ diamond line. Electrical measurements on films grown with 100% CH_4 and 1% CH_4 show the films to be quite conductive (10^{-1} Ω cm). However, films grown between 2 and 10% CH_4 are fairly resistive (10^{-7} Ω cm). Given that diamond-like carbon films can be quite resistive and given that these films show no 1332 cm⁻¹ line, it seems that the RPECVD under these conditions is producing a form of carbon similar to diamond-like carbon films typically deposited at much lower temperatures.

To investigate the possibility of heteroepitaxial growth, single crystal Ni(111) substrates were used. The crystals were obtained from Monocrystals, Inc.[9] unpolished. The samples were polished using a 0.1 μm colloidal silica polish. This avoids any embedment of diamond particles on the Ni surface which might serve as sites for homoepitaxial nucleation. Following an in-situ hydrogen clean, deposition began using 0.900 Torr, 200 sccm Ar (plasma), 100 sccm H_2 (ring), 1 sccm CH_4 (ring), and 100 W rf power. Figure 6 and 7 show scanning electron microscopy (SEM) photographs from one of the carbon layers. The surface showed shallow pyramidal structures whose faces were all aligned with one another. At higher magnification, one can see what appears to be growth fronts that were propagating across the Ni(111) surface. The dark trapezoids on the surface have been identified using Auger microprobe analysis as holes that have not overgrown. Note that the sides of all the holes are oriented with each other. This suggests that even the thin film is epitaxial in nature. The 4000 Å high pyramids with bases approximately 2 um wide were suggestive of {111} diamond planes extending from the Ni (111) substrate surface that had been truncated on the top. Like the polycrystalline samples, Raman analysis did not show a 1332 cm^{-1} diamond line but rather showed 1580, 1398, and 1350 cm^{-1} lines. Figure 8 shows a representative Raman spectra of the sample. Transmission electron microscopy (TEM) showed the sample not to be diamond but principally epitaxial graphite. Plan-view TEM specimens were produced by electropolishing the nickel substrate from the back side. Figure 9 shows a bright field micrograph accompanied by · selected area electron diffraction (SAD) patterns from one of the pyramidal structures and from the nickel substrate. The presence of the $01\bar{1}0$-type reflections from the hillocks are consistent with graphite being the predominant phase. The crystallographic orientation relationship observed between the nickel substrate and the graphite based hillock is:

$$[\bar{1}11]_{Ni} \mathbin{/\mkern-5mu/} [0001]_{C}$$
$$(101)_{Ni} \mathbin{/\mkern-5mu/} (1\bar{2}10)_{C}$$

and hence represents achievement of heteroepitaxial

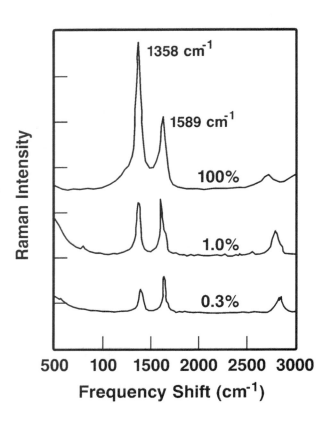

Figure 5. Raman spectra of carbon films on polycrystalline Ni substrates for different CH_4 concentrations

Figure 6. Low magnification SEM micrograph of expitaxial carbon on Ni (111)

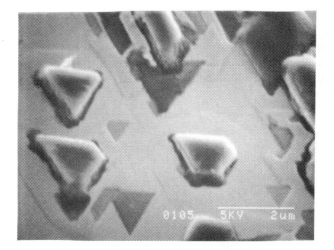

Figure 7. High magnification SEM micrograph of epitaxial carbon on Ni (111)

carbon on nickel on a microscopic scale. It is not possible with the current TEM to assess what epitaxial relationship exists between the Ni substrate and the thin film on the surface. Although we believe that the surface and gas phase chemistries will probably dominate the deposition process, it should be noted that there is a relatively small lattice mismatch between graphite and nickel ($d(220)_{Ni} = 1.246$ Å, $d(11\overline{2}0)_C = 1.231$ Å, mismatch = 1.2%).

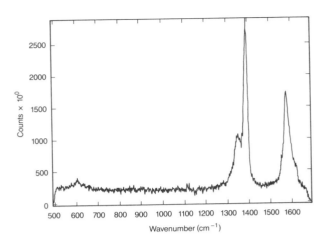

Figure 8.   Representative Raman spectrum from pyramidal sample

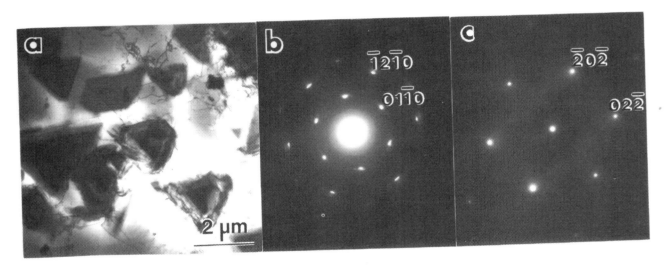

Figure 9.   Plan-view TEM micrographs showing pyramidal carbon hillock: (a) bright field, (b) [0001] graphite SAD pattern from hillock, (c) [$\overline{1}11$] SAD pattern from Ni. The depth of the pyramids are in <$\overline{1}10$>Ni type direction

3. CONCLUSIONS AND FUTURE DIRECTIONS

Remote plasma enhanced chemical vapor deposition has been used to investigate the nucleation of diamond and diamond-like carbon films on Ni surfaces. Our approach began by attempting growths under plasma conditions where metastable Ar atoms were the primary energy carrier entering the reaction zone. This resulted in minimal deposition. It seemed plausible that a rate limiting step such as H desorption from an H terminated surface was responsible for the lack of growth. Following this work, discharge conditions were changed such to allow other energy carriers into the reaction zone. Emission spectrum then showed evidence of CH species absent before and substantial C growth resulted. The layers were visibly opaque and electrically very conductive.

Hydrogen dilution experiments were then attempted in an effort to reduce the sp^2 bonding groups and enhance diamond growth. Work on polycrystalline Ni substrates showed the primary effects of hydrogen dilution under the unconfined plasma conditions were to reduce the growth rate and reduce the amount of disordered carbon as observed in Raman spectroscopy (i.e., a reduction in the 1350 cm^{-1} band). To investigate the possibility of heteroepitaxial growth, single crystal Ni (111) substrates were used. SEM photographs show the growth of oriented pyramidal features. Auger microprobe, Raman, and TEM plan-view analysis show the predominate phase to be graphite. Indeed, the crystallographic orientation observed between the Ni (111) substrate and the graphite based hillock represents achievement of heteroepitaxial carbon on nickel on a microscopic scale.

Future work on diamond heteroepitaxy will focus on developing CVD conditions under which diamond will be the predominate form. The authors suspect that this may take two directions. One direction would be a return to the confined discharge conditions where metastable activation is alone driving the reactions. For that approach to be successful, a means of inducing hydrogen desorption from that surface in a controlled manner must be developed. Another approach would be a continuation of the extended plasma work with emphasis on supplying more atomic hydrogen to the nucleating surface.

4. ACKNOWLEDGEMENTS

The authors would like to thank Bob Nemanich and Robert Schroder at North Carolina State University for Raman analysis on many of the samples. Special thanks also to Pehr Pehrson and Jim Butler at the Naval Research Laboratory for Auger Microprobe and additional Raman analysis. The work was supported under the Space Defense Initiative/Innovative Science and Technology Office through the Office of Naval Research Contract No. N-00014-86-C-0460

5. REFERENCES

1. C.P. Chang, D.L. Flamm, D.E. Ibbotson, and J.A. Mucha, "Diamond crystal growth by plasma chemical vapor deposition," J. Appl. Phys. 63(5), 1744-1748 (1988).

2. Seiichiro Matsumoto, Toyohiko Kobayashi, Mototsugu Hino, Takamasa Ishigaki, and Yusuke Moriyoshi, "Deposition of diamond in a rf induction plasma," in the International Symposium on Plasma Chemistry, Tokyo, 2458-2462 (1987).

3. Koji Kobashi, Kozo Nishimura, Yoshio Kawate, and Takefumi Horiuchi, "Summary Abstract: Morphology and growth of diamond films," J. Vac. Sci. Technol. A 6(3), 1816-1817 (1988).

4. D.J. Vitkavage, R.A. Rudder, G.G. Fountain, and R.J. Markunas, "Plasma enhanced chemical vapor deposition of polycrystalline diamond and diamond-like films," J. Vac. Sci. Technol. A 6(3), 1812-1815 (1988).

5. J. Balamuta, M.F. Golde, and Yueh-Se Ho, "Product Distributions in the reactions of excited noble-gas atoms with hydrogen-containing compounds," J.Chem.Phys. 79(6), 2822-2830 (1982).

6. R.C. Bolden, R.s. Hemsworth, M.J. Shaw, and N.D. Twiddy, "The measurement of Penning ionization cross sections for helium 2 3S metastables using a steady-state flowing afterglow method," J.Phys.B.: Atom. Molec. Phys. 3, 61-71 (1970).

7. D.W. Setser, D.H. Stedman, and J.A. Coxon, "Chemical applications of metastable argon atoms. IV. Excitation and relaxation of triplet status of N_2," J.Chem.Phys., 53(3), 1004-1021 (1970).

8. J.F. Prince, C.B. Collins, and W.W. Robertson, "Spectra excited in an argon afterglow," J.Chem.Phys., 40(9), 2619-2626 (1964).

9. Monocrystals, Inc., 1712-T Sherwood Boulevard, Cleveland, Ohio, 44117.

RAMAN ANALYSIS OF THE COMPOSITE STRUCTURES IN DIAMOND THIN FILMS

R.E. Shroder, R.J. Nemanich, and J.T. Glass

North Carolina State University, Departments of Physics and Materials Science and Engineering, Raleigh, NC 27695-8202

ABSTRACT

Raman spectroscopy has been used to examine diamond thin films produced by various CVD processes. The Raman spectra exhibit features which suggest that the films are composites of diamond (sp^3) and graphite-like (sp^2) bonding. A brief outline of Raman scattering from composites is presented. A first attempt at modeling these types of films using composites of diamond and graphite powders is reported. It is found that the Raman features associated with sp^2 bonding in the films do not correlate well to features exhibited by microcrystalline graphite.

1. INTRODUCTION

In recent years Raman scattering has been used to characterize the carbon films produced by various CVD processes[1-3]. This technique has proved useful for examining the diamond and graphite-like structures in the films. As the goal of the research is to produce single crystal, homoepitaxial diamond on a variety of substrates, Raman spectroscopy could be used to accurately determine the relative concentrations of these structures; thereby, demonstrating the quality of the materials produced in the deposition process. We have examined these possibilities using composites of graphite and diamond powders, and correlated the results to the diamond thin films.

The Raman spectra obtained from the carbon films have generally concentrated on the appearance of the first-order Raman mode of diamond at $1332 cm^{-1}$. Diamond is characterized by 4-fold coordinated sp^3-type bonding of F_{2g} symmetry[4]. It possesses a single triply degenerate zone-center optical phonon which corresponds to the highest energy mode in the phonon dispersion curves of diamond, so that features in the spectra with energy higher than $1332 cm^{-1}$ cannot be attributed to diamond structures of long range order. Other features do appear in the Raman spectra of carbon films from $\sim 1350 - 1600 cm^{-1}$, and are normally associated with structures having 3-fold coordinated sp^2-type bonding such as graphite. Films which display a variety of these features can be considered composites of sp^2 and sp^3 bonding (i.e. containing regions of differing atomic structures), and it is these types of films which have been closely examined. Particular attention has been focused on the relative intensities of the peaks associated with each type of bonding, in an effort to determine if the Raman scattering from diamond films can be used to give an estimate of the relative concentrations of sp^3 to sp^2-type bonding in the films.

The Raman spectra of two diamond films produced under differing deposition conditions are shown in Fig. 1. Both spectra contain the diamond peak at $1332 cm^{-1}$, but other features are present. Spectrum (a) also shows features at $1355 cm^{-1}$ and $1580 cm^{-1}$ which are similar to those associated with microcrystalline graphite[5], and a feature centered at $\sim 1500 cm^{-1}$ which has been assigned to regions of amorphous carbon.[6] Spectrum (b) shows a broad feature at $\sim 1500 cm^{-1}$ which is dominated by the diamond peak and has been ascribed to regions of highly disordered graphite[1]. In this study we will show that the size and absorption properties of these sp^2 and sp^3-type domains strongly affect the relative strengths of the Raman features described above. The details of Raman scattering from different carbon structures has been discussed elsewhere[6], so we provide a brief overview of Raman scattering from composites.

2. COMPOSITE PROPERTIES

The three types of composite structures considered are: micron-scale, microcrystalline, and atomically disordered. For micron-scale constituent materials, the Raman spectrum should appear as a linear combination of the features of the bulk materials. If one of the constituents of the composite is strongly absorbing, then light incident on particles of this type is completely absorbed. The effective absorption depth of the composite should be approximately equal to that of the highly absorbing constituent, modified by the relative concentrations of the absorbing and non-absorbing materials. As the domains decrease to microcrystalline scales, features should begin to appear in the Raman spectrum due to size effects associated with the breakdown of the conservation of wavevector[5]. The small domain size creates an uncertainty in the wavevector ($\Delta k \sim 2\pi/d$ where d is the domain size), and the momentum selection rules for the Raman scattering process are relaxed. The length scale of the crystallites is now on the order of or smaller than the absorption depths of the constituent materials and the decay length of the phonons, so that crystallite diffraction effects create a more uniform illumination of the composite. For an atomically disordered composite, the Raman spectrum becomes the average of the vibrational and electronic properties of the random network, and would appear quite different than the Raman spectra of the bulk materials.

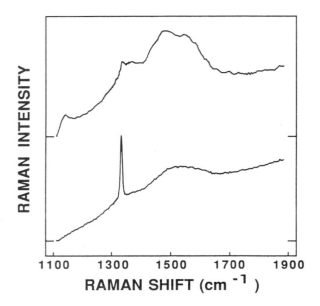

Fig. 1. First-order Raman spectra of diamond thin films. Both of the samples were produced by microwave plasma CVD. The sharp feature at 1322cm^{-1} is indicative of crystalline diamond while features between 1350 and 1600cm^{-1} are attributed to sp^2 bonded carbon.

3. EXPERIMENTAL

In order to model the ratio of sp^2 to sp^3-type bonding in diamond thin films, composites of diamond and graphite powders were examined. These powders were obtained from commercially available sources[7,8]. The diamond powder was slightly gray in appearance and of ~1μm crystallite size. The graphite powder was black and very cohesive and contained crystallites of 30-40μm size. Admittedly, this is large for our purposes, but we wished to examine the effects of the concentration of diamond in the composites. The examination of effects due to crystallite size was left for later. The composites were created as follows; appropriate amounts of diamond and graphite powder were weighed, and the relative concentrations determined. The powders were then mixed using mortar and pestle, and the resulting composites were lightly pressed into aluminum sample holders. The samples were mounted in the backscattering configuration, and the spectra were taken using the 5145Å line of an Ar⁺ laser.

4. RESULTS AND DISCUSSION

A. Raman spectra of diamond and graphite composites

Several of the Raman spectra taken from the composite samples are shown in Fig. 2. The two peaks exhibited are the first-order Raman mode of diamond at $1332 cm^{-1}$ and the first-order mode of graphite which has been shown previously to occur at $1580 cm^{-1}$ [9]. The relative concentration of diamond in the samples ranges from ~1% up to 60%. The most interesting feature of these spectra is that the ~1% diamond composite displays a 1:1 ratio between the peak intensities of the first-order modes of diamond and graphite. At 50% diamond, it is seen that the peak due to graphite has practically disappeared. Thus, the absorption of graphite has a significant effect on the Raman spectra of the composites, and is shown in the large disparity of the absolute intensity scales of the samples. As is to be expected, the relative intensities of the two peaks change in proportion to the concentration of diamond present in the composite. If the ratio of the peak intensities of diamond and graphite versus concentration of diamond in the samples is plotted, Fig. 3 is obtained. It is again noted that the diamond peak dominates the spectrum even at relatively low concentrations of diamond. The solid curve has been added as a guide to the eye, but the data might be fitted very well using an appropriate model of the Raman scattering from composites of diamond and graphite. This model could then be applied to the diamond films previously discussed and an estimate of the ratio of sp^2 to sp^3-type bonding obtained.

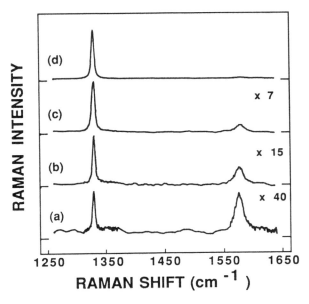

Fig. 2. First-order Raman spectra of the composites of diamond and graphite powders. The relative concentrations of diamond in the samples are: (a) 1.3% (b) 6.6% (c) 21.5% (d) 50.0% The spectra have been multiplied by the indicated value.

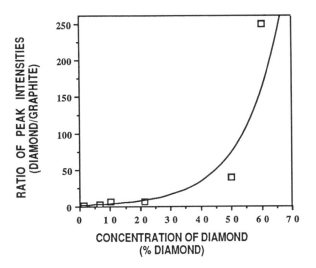

Fig. 3. Ratio of peak intensities (I_D/I_G) vs. relative concentration of diamond in the composite samples. The solid line is not a fit to the data.

B. Theoretical model

In order to model the diamond and graphite composites, we must understand the various parameters which effect the intensity of the Raman signal from these materials. Loudon[10] demonstrated that for a sample mounted in a backscattering geometry, the scattering intensity is given by

$$I = \frac{I_oS}{S+\alpha_1+\alpha_2}\{\ 1 - \exp[-(S+\alpha_1+\alpha_2)L]\ \} \tag{1}$$

where S is the scattering efficiency, I_o is the incident intensity, L is the sample thickness in the direction of the incident laser light, and α_1 and α_2 are the absorption coefficients at the respective frequencies of the incident and scattered light. Wada and Solin[11] showed that this equation could be modified to give the ratio of the scattering intensities of two different materials:

$$\frac{I_G}{I_D} = \frac{I_{oG}}{I_{oD}}\left(\frac{A_G}{A_D}\right)\left(\frac{1}{L_D(\alpha_1+\alpha_2)}\right)\left(\frac{\Delta\Omega_G}{\Delta\Omega_D}\right)\left(\frac{(1-R)^2}{(1-R_D)}\right)\left(\frac{(\sum_j[e_2\cdot\mathbf{R}_j\cdot e_1]^2)_G}{(\sum_j[e_2\cdot\mathbf{R}_j\cdot e_1]^2)_D}\right) \tag{2}$$

where S has been redefined in terms of a scattering efficiency, A, and a summation over the inner product of the Raman tensor, $\mathbf{R}_j$, of the degenerate first-order mode and the polarization unit vectors of the incident and scattered light, e_1 and e_2. I_G and I_D represent the scattered intensity from the graphite and diamond powders, and I_o represents the incident intensity . $\Delta\Omega$ is the solid angle into which light is scattered, and the term in R corrects for reflection of the scattered light at the sample surface and multiple reflections in the sample as in diamond. Here, α_1 and α_2 are the previously defined absorption coefficients of graphite since it has been assumed that diamond is transparent to the visible laser radiation.

If we now attempt to apply Eq. (2) to the composite samples, several difficulties must be faced. Since the Raman signal is being collected from a region of discreet particles, unique values for L_D and $\Delta\Omega$ no longer exist. Also, even if it assumed that each graphite particle is completely absorbing (i.e. much larger than the absorption depth), reflection losses due to light scattering between the diamond particles must still be accounted for. Finally, note that the last term in Eq. (2) provides unique values for the summations if it is known *a priori* the polarization directions of the incident and scattered light. Since the diamond and graphite composites contain particles of completely random orientation, an angle averaged value of the summations over all possible polarization directions must be taken.

We therefore modeled the diamond and graphite composites as follows; assuming that the collection angle ($\Delta\Omega$) is the same and neglecting reflective losses, the ratio of the scattering intensities of diamond to graphite may be written:

$$\frac{I_D}{I_G} \sim \frac{A'_D\ N_D\ V_D}{A'_G\ N_G\ V_G} \tag{3}$$

where A' is the angle and polarization averaged scattering efficiency, N is the atomic density of diamond or graphite, and V is the volume of material which is actually sampled by the Raman scattering. In terms of the bond densities of diamond and graphite, the percentage of diamond in the composite may be written as:

$$P_D = \frac{N_D}{\frac{3}{4}N_G + N_D} \quad \text{or re-writing} \quad \frac{N_D}{N_G} = \frac{3}{4}\left(\frac{P_D}{1-P_D}\right) \quad \text{so that Eq. (3) becomes}$$

$$\frac{I_D}{I_G} \sim \frac{3}{4}\frac{A'_D}{A'_G}\frac{V_D}{V_G}\left(\frac{P_D}{1-P_D}\right) \tag{4}$$

and the factor of 4/3 accounts for the fact that diamond contains 4 nearest neighbor bonds and graphite only 3 nearest neighborbonds per atom. The first-order Raman mode in both diamond and graphite is a bond stretching mode.

It has been reported[11] that the ratio between the scattering efficiencies of diamond and graphite is ~ 50:1, so that the only unknown in Eq. (4) is the ratio of the volumes, V_D/V_G. If we assume an average size for the graphite particles of ~ 30μm diameter, then we can model a unit volume of a 50% composite to be two spheres, each ~ 30μm diameter, one being graphite and the other, a loose collection of diamond particles. As was previously stated, the diamond is essentially transparent to the incident laser light, so that the entire "sphere" would be illuminated. At the 5145Å exciting wavelength, however, graphite has an absorption depth of ~ 300Å[11], although considerable discrepancies exist in the literature over the optical constants of graphite. Because the scattered light must also exit the absorbing region, we therefore consider the volume of graphite sampled to be a thin layer, ~150Å thick, across the top half of the graphite "sphere" as shown in Fig. 4. At P_D = 50%, our calculated value for I_D/I_G differs from that shown in Fig. 3 by a factor of two. This is in reasonable agreement considering our approximations.

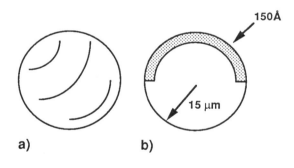

Fig. 4. Model of the unit volumes in the Raman scattering from composites. (a) fully illuminated diamond particles of ~ 30μm size (b) graphite particle of ~ 30μm size, partially illuminated in a 150 Å surface layer.

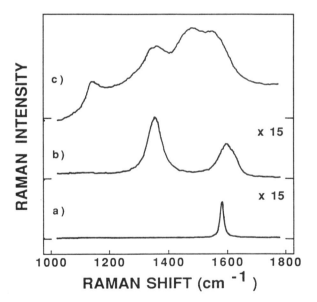

Fig. 5. First-order Raman spectra of (a) highly oriented pyrolytic graphite (b) "glassy" carbon (c) diamond thin film produced by microwave plasma CVD. The spectra have been multiplied by the indicated value.

We wish to emphasize the importance of crystallite size on our results. For smaller values for V_D and V_G, it is expected, from our model, that the peak in the Raman spectrum due to graphite would appear more strongly at higher concentrations of diamond than the results shown in Figs. 2, 3 indicate. This implies that the relative concentration of sp^3-type bonding in the films shown in Fig. 1 is higher than the composite samples suggest.

Upon examining the films in Fig. 1, we become aware of a very important characteristic of the features exhibited in the composite films. In Fig. 5, the Raman spectra of three materials containing sp^2-type bonding are shown. The highly oriented pyrolytic graphite (HOPG) and the "glassy" carbon samples both exhibit peaks at $1580cm^{-1}$ which have approximately the same absolute intensity. However, the composite diamond film shown in Fig. 5(c) has an absolute intensity ~15 times greater than these two. If the sp^2-type features in the diamond films are to be attributed to regions of crystalline or microcrystalline graphite, then this difference in the absolute intensity scales must be accounted for. Upon re-examination of Eq. (1), the two parameters which might significantly affect the intensity are the scattering cross section, A, and the absorption coefficients, α_1 and α_2. We propose that the regions of the diamond films attributed to graphite-like structures are actually regions of sp^2-type bonding, dissimilar to graphite, but distinctly different from a-C such that their respective features may be resolved in the Raman spectrum. The results presented in this paper suggest that the optical constants of these structures must be quite different from those of graphite.

4. Concluding Remarks

The Raman spectra of various diamond thin films have been examined. The features displayed by the films suggest that they are composites of sp^2 and sp^3-type bonding. We have attempted to understand the nature of these films using composites of diamond and graphite powder samples. We have proposed a simple model to the Raman scattering from our composites and found reasonable agreement to our experimental data. Our investigation demonstrates that both concentration and crystallite size have significant effects on the resulting Raman spectra. In diamond films containing regions of true diamond and graphite, we predict that the Raman spectra could be used to estimate the relative concentrations of these regions, but we emphasize that the sizes of the various domains in the films must first be known. We have found that the sp^2-type bonding in the films does not correlate well to graphite-like structures, and we therefore propose that the diamond films are composites of sp^3-type bonding having long-range order and sp^2-type bonding of some as yet unknown configuration.

5. Acknowledgement

We gratefully acknowledge K. Kobashi and Y. Kawate of Kobe Steel, Ltd, for supplying the diamond samples used in this study. We also appreciate the helpful discussions of R. Rudder at the Research Triangle Institute and B. Williams at North Carolina State University, as well as, the help of R. Russell at NCSU in the preparation of the composite powder samples. This work was supported in part by SDIO/IST through the Office of Naval Research under contract N00014-86-K-0666.

6. References

1. S. Matsumato, Y. Sato, M. Tsutsumi, N. Setaka, J. Mat. Sci. 17,3106(1982)
2. A. Sawabe and T. Inuzuka, Thin Solid Films 137, 89(1986)
3. K.Kurihara, K. Sasaki, M. Kawarada, N. Koshino, Appl. Phys. Lett. 52 (6), 437(1988)
4. S.A. Solin and A.K. Ramdas, Phys. Rev. B1, 1687(1970)
5. R.J. Nemanich and S.A. Solin, Phys. Rev. B20, 392(1979)
6. R.J. Nemanich, J.T. Glass, G. Lucovsky, R.E.Shroder, J. Vac. Sci. Tech. A6 (3), 1783(1988)
7. The diamond powder was obtained from AESAR CHEMICALS, Seabrook, NH 03874
8. The graphite powder was obtained from ALFA CHEMICALS, Danvers, MD, 01923
9. F. Tuinsta and J.L. Koenig, J of Chem. Phys. 53, 1126(1970)
10. R. Loudon, Adv. Phys. 13, 423(1964)
11. N. Wada and S.A. Solin, Physica 105B, 353(1981)

Raman spectroscopy of synthesized diamond grown by hot filament chemical vapor deposition

Edgar S. Etz

Gas and Particulate Science Division
Center for Analytical Chemistry

and

Edward N. Farabaugh, Albert Feldman, and Lawrence H. Robins
Ceramics Division
Institute for Materials Science and Engineering

National Institute of Standards and Technology
Gaithersburg, Maryland 20899

ABSTRACT

Raman microprobe studies of individual microcrystals of diamond and of thin diamond films deposited by the hot-filament chemical vapor deposition (CVD) method are focused on the determination of the purity of the diamond phase and on the extent and nature of defects of the diamond structure. The findings are discussed in relation to deposition parameters, growth mechanisms, and diamond morphology. The specimens consisted of single microparticles of sizes 3 to 40 μm, particle clusters, and continuous polycrystalline films of 3 to 8 μm thickness grown on silicon substrates. The interpretation of the results is based on the line shape, line width, and frequency position of the diamond line nominally at 1333 cm^{-1} Raman shift, as well as on other characteristic Raman bands in the region 1300 to 1600 cm^{-1} attributed to graphitic carbon components. Examined also are the relationship of the spectral background signal to the signal of the Raman features. Luminescence emissions arising from either structural imperfections or substitutional impurities in the diamond lattice are observed. A luminescence band centered around 738 nm (1.68 eV), attributed to either the neutral lattice vacancy in diamond, or possibly a silicon pair substitution in the diamond lattice, widely varied in intensity among the samples analyzed. The observation of this photoluminescence band is correlated with results from concurrent cathodoluminescence measurements.

1. INTRODUCTION

The widespread realization of diamond coatings by vapor phase synthesis has opened potential applications in numerous high-technology areas, and a considerable research effort has been placed on perfecting the methodologies for producing monocrystalline diamond films.[1-5] The diverse ways of depositing carbon films, or coatings, provide for a continuum of coatings with various structures including graphite, diamond-like carbon, diamond-like hydrogenated carbon, and diamond. Numerous methods have been utilized to characterize the properties of these coatings and films, and these investigations have all been essential to the understanding of nucleation and film growth. The current thrust of our research is directed at establishing the relationship between growth properties and the optoelectronic properties of diamond particles and diamond films produced by hot-filament chemical vapor deposition (CVD).

Raman microprobe spectroscopy provides several important advantages for the study of microcrystals of diamond and thin diamond films deposited by any of the new CVD techniques.[5-10] With the ability to focus the laser beam to a small probe spot, it becomes possible to measure the spectra of individual microcrystals or to probe the structure and composition of diamond films with a spatial resolution of several micrometers. With relatively good detection sensitivities for the relevant structures of interest, one can detect and identify many phases important to the characterization of synthesized diamond. Thus, in any analytical scheme for the comprehensive characterization of diamond films, micro-Raman spectroscopy will readily furnish a great deal of information from such specimens.

Of foremost interest is the identification of principal phases, diamond, graphite, and various forms of disordered and amorphous carbon, such as diamond-like carbon (DLC). The first-order Raman line of diamond is generally observed at 1333±1 cm^{-1}, and in the case of high-purity natural diamond, no other Raman features in the region 1300 to 1600 cm^{-1} are observed.[11] This spectrum has a negligible background signal, and, hence, virtually no luminescence. For CVD diamond, it becomes useful to examine the peak-to-background ratio at the Raman frequency of the diamond line as a measure of defects in the diamond phase. In addition, it is useful to note the exact frequency position of the diamond line as any frequency shift will be indicative of strain in the diamond lattice. Furthermore, the shape of the line is related to the distribution of distortions in the lattice both with regard to short range and long range order, and specifically the full width of the line at half the maximum signal (after subtraction of the background signal) is a relative measure of this distribution. In the spectral region of primary interest, the range 1300 to 1600 cm^{-1}, attention is focused on the presence of Raman lines indicative of graphite as a potential defect and impurity phase. The nucleation of graphitic carbon is nearly always in competition with that of diamond, and this process may result in the formation of a graphitic phase external to diamond crystals, such as at grain boundaries and at interfaces, but may also lead to sp^2 planar defects in the diamond matrix itself. Single-crystal graphite also shows one Raman-active mode of vibration and its Raman line, such as for highly-ordered pyrolytic graphite, is generally observed at 1575 cm^{-1} shift.[12,13] Microcrystalline graphite and various forms of disordered carbon usually show two more or less broad features centered at 1360 and 1600 cm^{-1} in the first-order spectra.[14-16] Of these, the 1360 cm^{-1} band in graphitic materials is usually assigned to a disorder mode which results from a size effect of grains in the graphite structure.[12]

Any spectral background level in the Raman spectra of diamonds arises from luminescence emissions by specific types of lattice defects. There is considerable interest in examining these photoluminescence (PL) emissions from diamond[17-20] and to correlate them with cathodoluminescence (CL) studies of synthetic diamond or diamond compacts.[21-25] In these investigations, as with the observation of PL spectra, the wide spectral regime from the near ultraviolet extending out to the red (i.e., the range from 225 nm to 1000 nm, corresponding to the photon energy range from 5.5 to 1.2 eV) is of interest, mainly because CL has proved a most successful technique for characterizing defects and impurities in diamond.

In this paper we describe the application of micro-Raman spectroscopy to gain information about localized atomic structure and composition and the microscopic spatial distribution of defects that may provide insight into the reasons for the polycrystalline growth habit of present-day films. Research described in a companion paper[7] provides details on the growth of diamond films examined in this work and discusses the various characterization techniques employed in our laboratories for the study of CVD diamond. In our evaluation of the micro-Raman spectra from these samples, we are progressively recognizing the information content inherent in these data. The laser-excited spectra are now routinely acquired over wide spectral ranges in the visible to also include potential luminescence emissions which, when present, are a direct reflection of optical transitions of defect-related electronic states in diamond. These same, or related, optical transitions have also been probed in parallel measurements by the application of CL imaging and spectroscopy on the same set of samples.[25] Thus, the CL emissions produced by electron beam excitation in the scanning electron microscope (SEM) provide complementary information to the laser-excited PL emissions recorded in the micro-Raman spectra, though in some cases different sets of defects may be excited in the two techniques. The Raman microprobe measurements have provided synergistic information in that for some samples they have revealed a PL band at 738 nm (1.68 eV) which appears to correlate well with a band at about the same energy in the CL emissions. Thus, these combined experimental techniques selectively examine specific defects that can have a pronounced influence on the optical and electronic properties of the diamond deposit.

2. EXPERIMENTAL

2.1 Deposition, Growth and Morphology of Diamond Film

The diamond particles and continuous films examined in this study were formed by hot-filament CVD, as described in greater detail in a companion paper.[7] The films were 3 to 8 μm thick. Substrates used were single-crystal silicon wafers and polycrystalline silicon carbide. The general growth conditions for diamond particle deposits and films were: substrate temperatures, 600 to 950 °C; deposition pressure, 40 torr (= 4×10^3 Pa); methane-to-hydrogen ratio, 0.005 by volume; tungsten filament temperature, 1800 °C. Quite generally the different CVD techniques produce diamond crystals and films with very similar morphologies.[3] There are kinetic factors common to all of these techniques, but other parameters such as plasma and surface chemistry are as yet poorly understood.

Figure 1 shows a typical deposition of diamond particles grown on a silicon wafer at 800 °C at a particle density not sufficient for continuous film formation. In general, the continuous films formed show a wide range of morphologies, but the particles often exhibit simple geometric shapes. They grow as cubes, octahedra, cubo-octahedra, single and multiple twins, polycrystalline layers, in rounded shapes, and in highly-irregular forms. At fixed conditions of composition, flow of reaction gas, and filament temperature, variations in the morphology of the surfaces comprising the continuous films are affected by changes in the substrate (i.e., the deposition) temperature. This is evident from a series of depositions on Si wafers where the substrate temperature was varied from 600 to 850 °C. The resulting films show a wide range of microstructure and complex morphological features,[7] and, from the results of our studies, could be inferred to possess various degrees of imperfections. It is known from examination of such films by transmission electron microscopy (TEM) that there are stress fields associated with the diamond/silicon interface.[2] Thus, CVD diamond is generally composed of identifiable single particles resulting in films that are polycrystalline.

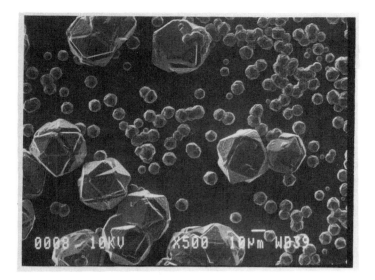

Figure 1. SEM micrograph of diamond microcrystals grown on (111) silicon substrate at 800 °C. The particles are twinned icosahedrons and combinations of (100) and (111) forms. Particle sizes in this deposition range from 4 to 40 μm.

2.2 Raman Microprobe Measurements and Photoluminescence

Two types of Raman microprobes were used for the spectroscopic characterization of these samples: (i) a scanning instrument employing a double grating spectrometer with single-channel (cooled photomultiplier tube) detection, and (ii) a multiplex-detection microprobe, employing a triple spectrograph with optical multichannel analyzer utilizing a cooled intensified diode array detector. Both microprobes are in-house designed and constructed systems employing as their major components standard commercial instrumentation and equipment. The fore-optical system, in each case, is centered on a microscope-type arrangement that allows the incident laser beam to be focused on the specimen and permits the collection of the scattered radiation in backscattering geometry. Both types of microprobes offer distinct and different advantages to the examination of microsamples which result from the different performance characteristics of the two types of Raman instruments. In both instruments, the microprobe spectra of the samples were excited with the 514.5 nm line of an argon-ion laser, employing a probe spot of several micrometers in diameter. Typical laser power levels focused into the probe spot ranged from 20 to 60 mW. In all cases the spectra were obtained at room temperature, and there was no indication of any laser-induced sample heating or modification from possible laser absorption effects by the sample. In the scanning microprobe, the spectra were acquired with 3 cm^{-1} spectral resolution. In the multichannel instrument, the spectral resolution is about 6 cm^{-1} for the spectral coverage selected and spectral range viewed by the diode array. In addition to observing the Raman emissions in these microprobe spectra, laser-excited PL emissions are observed from the samples whenever intrinsic defects, lattice distortions, or impurities give rise to such (visible) spectral emissions. To record these types of luminescent emissions, the spectra were generally acquired much beyond the normal Raman range (i.e., 0 to 3500 cm^{-1} from the exciting line). Specifically, the spectral range from 5000 to 6800 cm^{-1} (Raman shift from 514.5 nm), that is from about 690 to 790 nm, was examined for the presence of the characteristic luminescence band at 738 nm.

The effects of various deposition parameters, and their correlation with growth and morphology, were also studied concurrently by obtaining the CL spectra of these diamond films.[25] The cathodoluminescence was excited in these specimens by the electron beam of a scanning electron microscope. The CL spectra were observed over the wavelength range from about 400 to 850 nm (photon energy range from approximately 3.1 to 1.5 eV) with a conventional monochromator and diode array detector.

3. RESULTS AND DISCUSSION

The results reported here are those of work in progress to characterize synthesized diamond by various synergistic methods. Raman spectra are very sensitive to changes that disrupt the translational symmetry of solid phase materials, as occurs in small-dimensional crystals. The defects present in CVD diamond may be especially complex since there exist several types of bonding for elemental carbon (sp, sp^2, sp^3) depending on the thermodynamics and kinetics of the growth conditions. These relationships may produce material with a broad range in structure, such as the materials commonly referred to as "diamond-like carbon" (DLC) which are primarily amorphous with variable sp^2 to sp^3 ratios and whose composition is of widely varying hydrogen content.[26-29] The films generally considered "pure" diamond are structures stabilized by sp^3 carbon atoms, this being achieved by the chemistry in the gas or plasma and at the surface of the growing film. A continuing goal of our research is to perfect the deposition conditions which will produce films with a high degree of sp^3 bonding and thereby come close to the properties of single-crystal diamond. The Raman microprobe measurements provide insights about particle nucleation and film growth by obtaining spatially resolved information about microstructure, grain boundaries, and interfaces. The results reported here demonstrate the usefulness of this approach. The results are by no means complete since this work has been limited to a specific set of samples.

The early depositions on silicon, and other substrates, resulted in the growth of individual diamond particles only.[5,7] These isolated, single particles (size range < 1 to 15 μm) were ideal for examination in the Raman microprobe. Particles down to 2 μm in size could be interrogated, furnishing strong Raman signals. However, because of the small total quantity of material, no supporting x-ray diffraction data could be obtained. Numerous particles in these deposits furnished spectra in which the frequency of the diamond line was lower by 8 to 10 cm^{-1} from the "reference" Raman shift of 1333±1 cm^{-1} for bulk, single-crystal diamond.[11] For these particles, the diamond line was observed on a very low spectral background. The measured half-width (FWHM) of 4 to 5 cm^{-1}, typically, compared well with the experimentally observed half-width of 3 cm^{-1} from gem-quality diamond. We have tentatively attributed this shift in the Raman frequency to stress effects (calculated to amount to 3×10^9 Pa isostatic tensile stress),[5,7] and are continuing to examine samples of this type for this apparent anomaly in the frequency position of the diamond line. More recent measurements on several depositions of diamond particles on silicon wafer substrates (of both (111) and (100) orientation) have produced consistent spectra, though with wide differences in the level of the spectral background. Representative of these results is the spectrum shown in Figure 2 obtained from a single icosahedral microcrystal ≈12 μm in size. This spectrum was scanned over two spectral regions, i.e., the normal Raman range to record the Raman emissions from the substrate and those of the sample, and the second range, to higher Raman shifts, to detect the presence of any narrow-band luminescent emissions over the range from 690 to 790 nm. In this and the other particle spectra, the diamond line maximizes at 1334 cm^{-1}, consistent with the literature data. Also, the principal silicon phonon, a convenient internal calibration line at 522 cm^{-1}, appears at the expected frequency. The scattering from the silicon substrate does not present an interference in these spectra from diamond. The half-width of the diamond line is 6 cm^{-1}, the line appearing on a moderate spectral background. Because of the absence of any spectral features in the graphitic region from 1500 to 1600 cm^{-1}, it can be concluded that there is no graphitic carbon component present and that this diamond is nearly free of any defects or impurities. Examining the second spectral range, one finds a weak, broad feature centered at 5890 cm^{-1} (1.68 eV) which we have identified as the 738 nm luminescence band, subsequently observed in numerous other spectra of single particles and films.

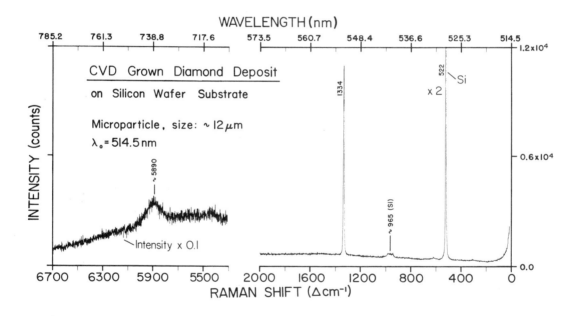

Figure 2. Representative micro-Raman spectrum of a single diamond particle of a deposition at 800 °C on (111) Si wafer. Spectrum obtained with the scanning microprobe, over two spectral regions. Raman bands from the silicon substrate are marked.

The spectrum shown in Figure 3 was also obtained with the scanning microprobe. It is that of a thin (≈8 μm thickness) continuous diamond film deposited at 700 °C on (111) silicon. This spectrum is significantly different from those of single microparticles examined in several sparse depositions on silicon (c.f. Fig.2). One obvious feature is the relatively higher spectral background across the full Raman range and extending out to the region of the luminescence band centered at 738 nm. Several spectra were obtained from arbitrary probe spots on this film. In all cases, the diamond line peaks at 1334±2 cm^{-1} and its half-width varies from 10 to 14 cm^{-1}, thus it is approximately twice as broad as the line observed for the more perfect diamond microcrystallites. The diamond line is superimposed on a rising background which, to lower energies, shows a broad band centered at 1510 cm^{-1}. This peak is attributed to diamond-like carbon (DLC). This band on the high frequency side of the diamond line can be related to planar defects from the incorporation of sp^2 bonds into the diamond lattice. It is more pronounced in the spectra of disordered phases; these specimens still show x-ray diffraction patterns characteristic of pure diamond. Thus, in this film and other films like

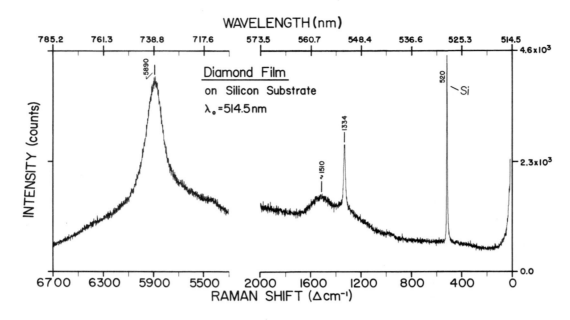

Figure 3. Micro-Raman spectrum of a continuous diamond film (thickness, ≈8 μm) deposited on (111) Si substrate at 700 °C. Spectrum obtained with the scanning microprobe. Film shows pronounced luminescence emission with peak at ≈738 nm (1.68 eV).

it, there appear to be varying mixed phases of tetrahedral (sp^3 bonding) diamond and DLC or graphitic carbon (sp^2-type bonding). In the spectra of these films, the luminescence gradually increases and extends out to beyond 5000 cm^{-1} where the luminescence develops into a distinct and pronounced band centered at 5890 cm^{-1}, the 738 nm photoluminescence peak. The luminescence emissions, over this spectral range, are also observed in the CL spectra from these samples.[25] The results indicate, quite generally, that the thin polycrystalline film deposits are cubic diamond but appear to have some defects as compared with natural diamond.

The spectra shown in Figures 4 to 6 were obtained with the multichannel-detection Raman microprobe. In these cases, the spectrograph was set to pass the indicated two spectral ranges on to the diode array detector, in two successive measurements. Due to the performance characteristics of this type of detector (cooled to -20 °C), the spectra appear "different" from those obtained in the scanning microprobe due to a characteristic detector background which has not been subtracted out. Yet, these multichannel spectra, acquired over a 3-min integration time, show all of the important features consistent with the data obtained from the scanning instrument. In this series of measurements, we have examined thin films and single diamond particles, mainly on silicon substrates (Figs. 4 and 5) and on polycrystalline SiC (Fig. 6). It is these recent results which have been most extensively corroborated by parallel CL measurements on the same set of samples.

The spectra shown in Figure 4 are of three samples of diamond films on (100) Si deposited at 650, 750, and 850 °C. The first two films show Raman and luminescence patterns typical of the spectra of the set of films prepared at temperatures between 600 and 800 °C. The diamond line appears on a rising spectral background; its peak maximizes at 1332±2 cm^{-1}, and in all cases the spectra show a broad sp^2 impurity band that peaks in the region 1530 to 1560 cm^{-1}. This is the band again which reflects in-plane defects between the basic structural units of the diamond phase. Toward higher wavenumbers, the luminescence background in these spectra progressively increases and, at the onset of the second spectral range displayed, this luminescence comes down as it descends from a broad luminescence band extending from the yellow to the red (550 to 750 nm). This laser-excited PL band is also observed in the CL spectra of the same specimens. In the film deposited at 850 °C, the luminescence is of a different nature. Instead of the broad PL/CL band observed in this region of the spectra, the spectral background over the range 700 to 760 nm is relatively low, except for the fairly narrow luminescence band with maximum at 738 nm (5890 cm^{-1} Raman shift) which is quite pronounced.

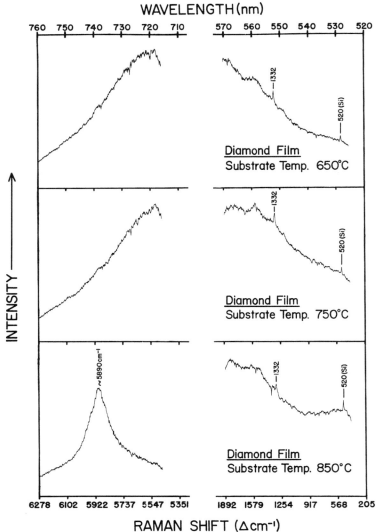

Figure 4. Raman spectra acquired with the multichannel detection Raman microprobe of three specimens of diamond film deposited on (100) Si substrates. Spectra obtained from films deposited at substrate temperatures of 650, 750, 850 °C. Raman and luminescence emissions shown are over two spectral regions. Only the film deposited at 850 °C shows the strong luminsence band at ~738 nm (1.68 eV).

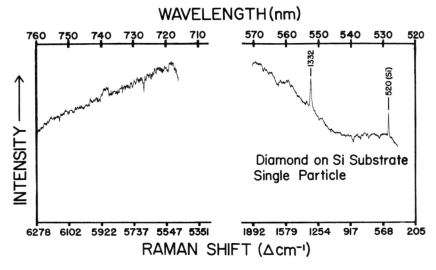

Figure 5. Raman spectrum acquired with the multichannel microprobe of a single diamond particle (size, ~35 μm; c.f. Fig. 1) in a deposition at 800 °C on (111) Si substrate. Higher wavelength spectral region shows only hint of the luminescence band at ~738 nm.

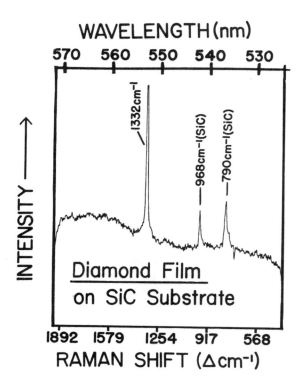

WAVELENGTH(nm)

Figure 6. Micro-Raman spectrum, obtained with the multichannel microprobe, of a continuous diamond film (thickness, ≈6 μm) deposited at 800 °C on polycrystalline silicon carbide. The principal Raman bands of the SiC substrate are indicated.

The spectra shown in Figure 4 are of three samples of diamond films on (100) Si deposited at 650, 750, and 850 °C. The first two films show Raman and luminescence patterns typical of the spectra of the set of films prepared at temperatures between 600 and 800 °C. The diamond line appears on a rising spectral background; its peak maximizes at 1332±2 cm⁻¹, and in all cases the spectra show a broad sp^2 impurity band that peaks in the region 1530 to 1560 cm⁻¹. This is the band again which reflects in-plane defects between the basic structural units of the diamond phase. Toward higher wavenumbers, the luminescence background in these spectra progressively increases and, at the onset of the second spectral range displayed, this luminescence comes down as it descends from a broad luminescence band extending from the yellow to the red (550 to 750 nm). This laser-excited PL band is also observed in the CL spectra of the same specimens. In the film deposited at 850 °C, the luminescence is of a different nature. Instead of the broad PL/CL band observed in this region of the spectra, the spectral background over the range 700 to 760 nm is relatively low, except for the fairly narrow luminescence band with maximum at 738 nm (5890 cm⁻¹ Raman shift) which is quite pronounced.

The spectrum shown in Figure 5 is that of a single particle of diamond deposited on (111) Si, at a deposition temperature of 800 °C. It is from one of the larger particles (size, ≈35 μm) shown in the SEM micrograph of Figure 1. The spectrum displays the basic features of the film spectra shown in Figure 4. The diamond line, positioned on a rising background, is a bit more pronounced,

The spectrum shown in Figure 5 is that of a single particle of diamond deposited on (111) Si, at a deposition temperature of 800 °C. It is from one of the larger particles (size, ≈35 μm) shown in the SEM micrograph of Figure 1. The spectrum displays the basic features of the film spectra shown in Figure 4. The diamond line, positioned on a rising background, is a bit more pronounced, and the graphitic carbon band is present. In the second spectral region, the 738 nm luminescence band shows up as a weak feature upon a descending spectral background.

It is now appropriate to consider the nature of the narrow luminescence band peaking at 738 nm (1.68 eV) which is observed in the spectra of single particles and films. In previous discussions of our micro-Raman results from CVD diamond,[5,7] we have tentatively interpreted this band as being of the same or of similar origin to a band in irradiated natural diamond which has been explained by isolated neutral vacancies.[17,18] Other data in the recent literature suggest that the appearance of this band may be related to the incorporation of silicon (an isoelectronic impurity) into the diamond lattice.[21,22] A cathodoluminescence line at about the same photon energy, namely at 1.685 eV, has been observed in silicon-implanted diamond. This luminescence associated with silicon impurities is believed to originate from a two-silicon center.[22] In the cathodoluminescence spectra of our films, we observe a band at 1.675 eV in the film deposited at 850 °C. The general photoluminescence properties of these films have been explained in part by internal strain effects in crystal grains and by vacancies in the diamond lattice that are trapped by nitrogen defects to form luminescence centers.[19,23]

Of the micro-Raman measurements on diamond films deposited on substrates other than silicon, one result is shown in Figure 6. It is a representative spectrum of a continuous film deposited at 800 °C on polycrystalline silicon carbide. Under identical process conditions, we observed the film growth on SiC to be faster than on Si, but the films exhibited quite similar microstructure and morphology.[5,7] The spectra from several probe spots on this film were consistent in that they all showed a moderate spectral background, thereby exhibiting a favorable peak-to-background relationship with respect to the diamond line. The frequency position and half-width (FWHM = 14 cm^{-1}) of the line point to a relatively defect-free pure phase of diamond also born out by the absence of any sp^2-bonding Raman feature in the 1530 to 1560 cm^{-1} region. Examination of the second spectral range (not presented in Figure 6) also showed no narrow-band luminescence feature at 738 nm. The diamond film on silicon carbide contains larger grains than has been found on silicon, which may explain why the growth on SiC has produced a film of a lower defect (e.g., twinning) density and a lower concentration of impurity atoms. These and other factors need to be clarified in future work.

4. SUMMARY

Micro-Raman spectroscopy has been applied to the study of CVD diamond deposits to examine the relationship between lattice defects and impurities and the crystallization process of synthetic diamond. We have observed a broad variety of spectra, from very close to single-crystal, natural diamond to those representing highly disordered phases. The results show a direct link between the growth features and defects observed in deposited diamond crystals. The spectra of the samples showing disordered phases imply the presence of an sp^2 bonding network mixed with an sp^3 bonding matrix, which can be explained by the incorporation of a graphitic carbon phase in the diamond lattice. These lattice defects are a reflection of imperfect growth leading to the formation of polycrystalline films. The broadening of the Raman line of diamond observed for films can be related to atom displacements a graphitic carbon phase in the diamond lattice. These lattice defects are a reflection of imperfect growth leading to the formation of polycrystalline films. The broadening of the Raman line of diamond observed for films can be related to atom displacements in the crystal lattice, resulting in a distortion of the long-range order of the lattice. In the spectra of many samples, diagnostic luminescence emissions have been observed of which a PL band at 1.68 eV is of principal interest. This band has been correlated with the corresponding CL band observed at the same photon energy for these samples. We relate this PL/CL band to the neutral lattice vacancy in diamond, but more recent insights concerning its origin point to defect structures that may result from a silicon atom dispersion in the diamond lattice. The latter is more probable in the microwave-assisted CVD of diamond where etching of the reaction chamber silica liner by the plasma is possible.

5. REFERENCES

1. R. Messier, A.R. Badzian, T. Badzian, K.E. Spear, P. Bachman, and R. Roy, "From diamond-like carbon to diamond coatings," Thin Solid Films 153, 1-9 (1987).
2. H. Kawarada, K.-S. Mar, J. Suzuki, T. Ito, H. Mori, H. Fujita, and H. Hiraki, "Characterization of diamond particles and films by plasma-assisted chemical vapor deposition using high-voltage electron microscopy," Jpn. J. Appl. Phys. 26, L1903-L1906 (1987).
3. A.R. Badzian and R.C. deVries, "Crystallization of diamond from the gas phase," Mater. Res. Bull. 23, 385-400 (1988).4. A.R. Badzian, T. Badzian, R. Messier, K.E. Spear, and R. Roy, "Crystallization of diamond crystals and films by microwave-assisted CVD," Mater. Res. Bull. 23, 531-548 (1988).
5. A. Feldman, E.N. Farabaugh, Y.N. Sun, and E.S. Etz, "Diamond, a potentially new optical coating material," in Laser Induced Damage in Optical Materials, NIST Special Publication, in press (1988).
6. L.S. Plano and F. Adar, "Raman spectroscopy of polycrystalline diamond films," Proc. SPIE Conf. 822, (1987).
7. E.N. Farabaugh, A. Feldman, L.H. Robins, and E.S. Etz, "Growth of diamond films by hot filament chemical vapor deposition," Proc. SPIE Conf. 969, (1988).
8. M. Ramsteiner, J. Wagner, Ch. Wild, and P. Koidl, "Raman scattering from extremely hard amorphous carbon films," J. Appl. Phys. 62, 729-731 (1987).
9. A. Sawabe and T. Inuzuka, "Growth of diamond thin films by electron-assisted chemical vapor deposition and their characterization," Thin Solid Films 137, 89-99 (1986).
10. R.O. Dillon, J.A. Woollam, and V. Katkanant, "Use of Raman scattering to investigate disorder and crystallite formation in as-deposited and annealed carbon films," Phys. Rev. B29, 3482-3489 (1984).
11. S.A. Solin and A.K. Ramdas, "Raman spectrum of diamond," Phys. Rev. B1, 1687-1698 (1970).
12. F. Tuinstra and J.L. Koenig, "Raman spectrum of graphite," J. Chem. Phys. 53, 1126-1130 (1970).
13. M. Nakamizo, H. Honda, and M. Inagaki, "Raman spectra of ground natural graphite," Carbon 16, 281-283 (1978).
14. P. Lespade, R. Al-Jishi, and M. Dresselhaus, "Model for Raman scattering from incompletely graphitized carbons," Carbon 20, 427-431 (1982).
15. J.N. Rouzaud, O. Oberlin, and C. Beny-Bassez, "Carbon films: structure and microtexture (optical and electron microscopy, Raman spectroscopy)," Thin Solid Films 105, 75-96 (1983).
16. C. Beny-Bassez and J.N. Rouzaud, "Characterization of carbonaceous materials by correlated electron and optical microscopy and Raman microspectroscopy," Scanning Electron Microscopy/1985/I, 119-132 (1985).
17. J.E. Field, Edit., The Properties of Diamond, Academic Press, London (1979).
18. J. Walker, "Optical absorption and luminescence in diamond," Rep. Prog. Phys. 42, 1605-1659 (1979).

19. T. Evans, S.T. Davey, and S.H. Robertson, "Photoluminescence studies of sintered diamond compacts," J. Mater. Sci. 19, 2405-2414 (1984).

20. J. Wagner and P. Lautenschlager, "Hard amorphous carbon studied by ellipsometry and photoluminescence," J. Appl. Phys. 59, 2044-2047 (1986).

21. V.S. Varilov, A.A. Gippius, A.M. Zaitsev, B.V. Deryaguin, B.V. Spitsin, and A.E. Aleksenko, "Investigation of the cathodoluminescence of epitaxial diamond films," Sov. Phys. Semicond. 14, 1078-1079 (1980).

22. A.M. Zaitsev, V.S. Varilov, and A.A. Gippius, "Cathodoluminescence of diamond associated with silicon impurity," Soviet Physics - Lebedev Institute Reports, pp. 15-17 (1981).

23. A.T. Collins and S.H. Robertson, "Cathodoluminescence studies of sintered diamond," J. Mater. Sci. Lett. 4, 681-684 (1985).

24. H. Karawada, K. Nishimura, T. Ito, J. Suzuki, K.-S. Mar, Y. Yokota, and A. Hiraki, "Blue and green cathodoluminescence of synthesized diamond films formed by plasma-assisted chemical vapor deposition," Jpn. J. Appl. Phys. 27, L683-L686 (1988).

25. L.H. Robins, L.P. Cook, A. Feldman, and E.N. Farabaugh, "Cathodoluminescence of defect levels in thin diamond films grown by hot-filament chemical vapor deposition," submitted to Phys. Rev. B.

26. S. Matsumoto, Y. Sato, M. Tsutsumi, and S. Setaka, "Growth of diamond particles from methane-hydrogen gas," J. Mater. Sci. 17, 3106-3112 (1982).

27. J.C. Angus, J.E. Stultz, P.J. Schiller, J.R. MacDonald, M.J. Mirtlich, and S. Domitz, "Composition and properties of the so-called 'diamond-like' amorphous carbon films," Thin Solid Films 118, 311-380 (1984).

28. C.B. Zarowin, N. Venkataramanan, and R.R. Poole, "New 'diamondlike carbon' film deposition using plasma assisted chemical vapor transport," Appl. Phys. Lett. 48, 759-761 (1986).

29. D. Nir, "Intrinsic stress in diamond-like carbon films and its dependence on deposition parameters," Thin Solid Films 146, 27-43 (1987).

Characterization of diamond-like films

N.R. Parikh, W.K. Chu, G.S. Sandhu, M.L. Swanson,
C. Childs, J.M. Mikrut and L.E. McNeil

Department of Physics and Astronomy, University of North Carolina
Chapel Hill, North Carolina 27599-3255

ABSTRACT

Thin diamond-like and diamond films, grown by remote plasma-enhanced CVD (RPECVD) and by plasma CVD, were characterized using optical microscopy, elastic recoil detection spectroscopy (ERD), and Raman scattering. The H concentration of the films was measured by ERD and was related to the growth parameters and to the quality of the films as determined by Raman scattering. The H content of samples grown by RPECVD at the Research Triangle Institute (RTI) increased with decreasing growth temperature, varying from 6 at% H at growth temperatures from 500-720°C, to 25 at% H at a growth temperature of 20°C. The characteristic Raman frequency of natural diamond, 1332 cm^{-1}, was observed for samples grown by Crystallume Corp. and by General Electric Corp. The samples obtained from Crystallume showed circular "bull's-eye" features by optical microscopy, and the width of the 1332 cm^{-1} peak was broad, indicating highly strained crystallites. For a sample from G.E., most of the film was good quality diamond, as indicated by a sharp 1332 cm^{-1} diamond frequency; the bottom (substrate) part of the film contained more H than the top part.

1. INTRODUCTION

Rapid progress has been made in the growth of diamond and diamond-like films by deposition from the vapor phase, using both plasma CVD[1-4] and remote plasma-enhanced CVD[5]. Of special importance for continuing progress in this field is the characterization of diamond films. In particular, the impurity content of the films is critical, as it can affect both the film integrity and its properties. Since most films are prepared using a CH_4-H_2 mixture, the hydrogen content of the films is especially important. We have measured the H concentration of CVD diamond-like films using an ion beam method, elastic recoil detection spectroscopy (ERD). In addition, we have used Raman spectroscopy in a microbeam mode to specify the quality of the diamond-like films as a function of depth, and to relate the film quality to the H content. Ellipsometry (not reported here) and optical microscopy have also been used to characterize the films.

2. EXPERIMENTAL PROCEDURE

Samples of diamond or diamond-like films were obtained from RTI, Crystallume and G.E. The films from RTI were grown on Si single crystals by remote plasma enhanced CVD, and the other films were grown by plasma CVD. These samples are not necessarily state-of-the-art or the best films produced by these suppliers; they are however representative examples for demonstration of our characterization methods. The samples were examined by optical and scanning electron microscopy, elastic recoil detection spectroscopy, Raman spectroscopy and ellipsometry.

2.1 Elastic recoil detection spectroscopy (ERD)

ERD is a fast and non-destructive method[6] of obtaining H profiles up to depths of about 600 nm. By this method, a beam of MeV energy light ions, for example He+, is directed at the target, causing a small fraction of the H atoms in the target to be ejected by elastic recoil collisions. The concentration of H atoms versus depth in the target can be calculated from the observed energy spectra of the recoiled H atoms, using standard kinematic analysis. The parameters required are the initial energy E_O of the ion beam, the He-H Rutherford collision cross section σ, the detector solid angle Ω, the scattering angle of the recoiled H ions, the energy distribution of the H ions, and the stopping powers S_1 of the incident He ions and S_2

of the recoiled H ions in the target material. Often a standard material containing a known amount of H is used instead of measuring σ and Ω.

The concentration C_H of H atoms in the sample is given by

$$C_H = N(E_H)dE_H/(I\Omega\sigma(E_1)dx), \qquad (1)$$

where $N(E_H)$ is the measured number of recoiled H ions per unit energy, dE_H is the energy increment corresponding to a depth increment dx, I is the number of incident ions, Ω is the detector solid angle and $\sigma(E_1)$ is the cross section for Rutherford scattering at the incident He energy E_1.

This technique has a sensitivity of about 0.01 at% H and a depth resolution of about 20 nm. In the present experiments, a beam of 2 MeV ^{4}He+ was used, at an incident angle of 15° from the surface, and the recoiled H+ ions were detected at a forward scattering angle of 30°, as shown in the inset of Fig. 1. The energies of the recoiled H ions were determined by a surface barrier detector which was covered with a 6μm thick Mylar film to shield the backscattered He+ ions.

2.2 Raman Scattering

Raman scattering[7,8] can be used for nondestructive evaluation of the quality of a diamond film, as the dominant zone-center vibrational mode frequencies of diamond (1332 cm^{-1}), graphite (1589 cm^{-1}) and disordered carbon (1360 cm^{-1}) are easily distinguished. The scattering volume examined depends strongly on the penetration depth of the incident light. For incident wavelengths at which the material is transparent the entire thickness can be sampled, but for wavelengths at which it is opaque only the near-surface region can be examined.

Raman scattering measurements were performed using a 0.8 m double-grating spectrometer equipped with photomultiplier detection and photon-counting electronics. Illumination was provided by the

5145 Å line of an Ar$^+$ laser. The spectral resolution was 1.2 cm^{-1} and the polarization of the scattered light was not analyzed.

For coarse measurements conventional focussing optics were used to produce a laser spot size of ~ 150 μm on the sample. For detailed examination a microscope was used to focus the laser light to a 1.6 μm spot so that features on the sample could be examined at that spatial resolution.

3. RESULTS

3.1 RTI Samples

ERD measurements of a diamond-like film grown at 400°C from RTI are shown in Fig. 1. Spectra are given for two different spots on the sample; the different energy widths indicate a 20% variation in film thickness for the two spots. The H concentrations were almost identical for the two spots, as shown by the equal numbers of H counts (the detected number of H recoils). The H concentration profile was obtained directly from the number of H counts versus energy (which is directly proportional to the ERD channel number on the figure), using equation (1). A Mylar film was used as a standard. The H composition was approximately 16 at%, and was almost constant through the film thickness. For RTI films grown at different temperatures, the amount of H in the RTI films decreased with increasing growth temperature, varying from 25 at% for 20°C growth to 6 at% for 500-720°C growth (Table 1).

3.2 Crystallume Samples

ERD results for the Crystallume samples are shown in Fig 2. The films were somewhat thinner than the RTI ones. The H contents for different films are shown in the figure, and varied from 8 at% H to 15 at% H.

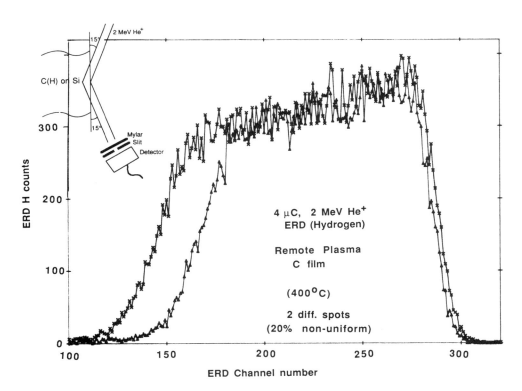

Fig. 1 Elastic recoil spectra recorded at 293K for two different spots on a RTI diamond-like C film, deposited by RPECVD at 400°C on a Si substrate. The incident He[+] energy was 2 MeV. The counted number of recoiled H[+] ions is plotted as a function of their energy (ERD channel number). The inset shows the sample vs detector configuration.

TABLE I

RESULTS OF ERD ANALYSIS OF C(H) FILMS GROWN BY REMOTE PLASMA CVD

Sample Growth Temp. (°C)	C + H (10^{18}/cm^2)	Layer Thickness (nm) assuming 10^{23} atoms/cm^3	H/(H+C) (at%)
20	10	100	25
400	4.0	40	16
500	3.3	33	6
600	0.7	7	
650			7.5
720			4.2

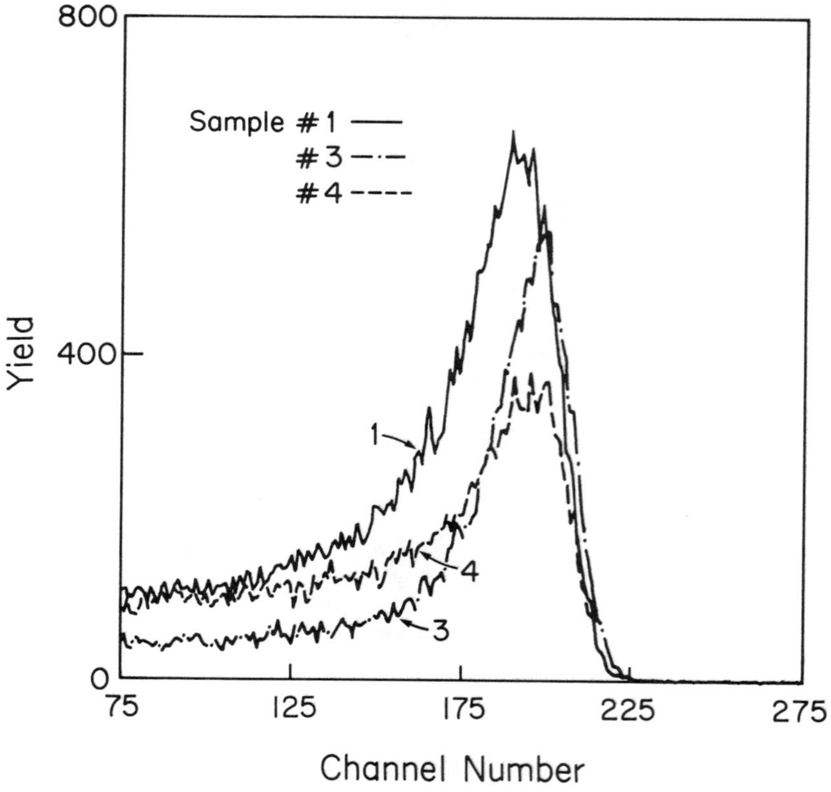

Fig. 2.   Elastic recoil
spectra for Crystallume
samples, recorded as in
Fig. 1. The three spectra
were obtained for different
samples.

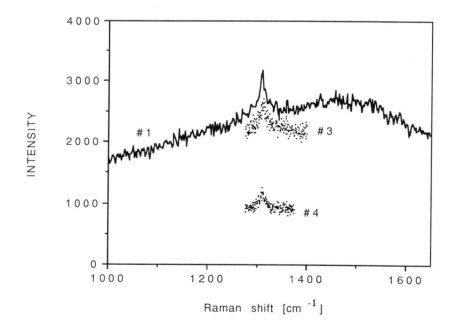

Fig. 3. Raman scat-
tering spectra for
three different
Crystallume samples
using a 150 μm beam
spot. The charac-
teristic diamond
frequency at 1332
cm^{-1} is visible
above a broad
background

Raman scattering results for these samples showed the presence of
the characteristic diamond frequency at 1332 cm^{-1}, as illustrated in
Figure 3.   These results were obtained with low spatial resolution

(150μm spot), which in this case produced a broad asymmetric background peak, which is probably fluorescence due to an impurity. Microscopic examination showed that these samples were highly nonuniform. All three samples consisted of a uniform background (which showed no indication of diamond), spotted with a series of roughly circular "bull's eyes" with a visible ring structure (Fig.4). On sample #1 these features were typically ~ 100 μm in diameter and were few in number. On the other two samples the "bull's eyes" were greater in number but smaller in diameter, typically 50 - 60 μm. The "bull's eyes" apparently represent nucleation sites, as they showed strong diamond vibrations; high spatial resolution Raman spectra of all points examined on these features showed a peak at 1332 cm^{-1} with FWHM of 9 - 13 cm^{-1}, with no evidence of disordered carbon. The large widths recorded suggest that the diamond crystallites were small and rather strained.

3.3 General Electric Samples

The GE samples showed good quality diamond when examined by Raman scattering on both the top and the bottom (substrate side) surfaces. The average peak position was 1331.2 cm^{-1} and the full width at half maximum (FWHM) varied from 3.9 to 5.1 cm^{-1}, indicating that the examined parts of the film were of good crystal quality but not single crystals. These samples were cut by a laser beam from larger pieces. When a film was examined on the edges by optical reflectance microscopy, distinct layers were seen, as shown in Fig. 5 (right side). These layers appeared to be due to the effect of the laser cutting, since no such variation through the thickness of the film was observed at a cleaved edge (left side of Fig. 5). In confirmation of this observation, Raman scattering data obtained with the microbeam showed a strong diamond peak at 1332 cm^{-1} for all depths on the cleaved edge (darkly circled spots on Fig. 5). However, for the laser-cut edge, only the top surface showed the strong diamond signal (Fig. 6, top spectrum), whereas the region near the bottom surface showed only a weak disordered carbon signal (Fig. 6, bottom spectrum).

Elastic recoil measurements taken on each surface of the G.E.

sample are shown in Fig. 7. The region near the top surface had a H concentration only equivalent to that of hydrocarbon contamination on the surface, whereas the region near the bottom surface contained 6 at% H.

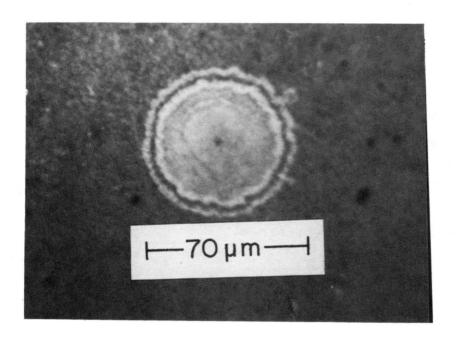

Fig. 4. Optical micrograph showing "bull's eye" ring structure on a Crystallume film.

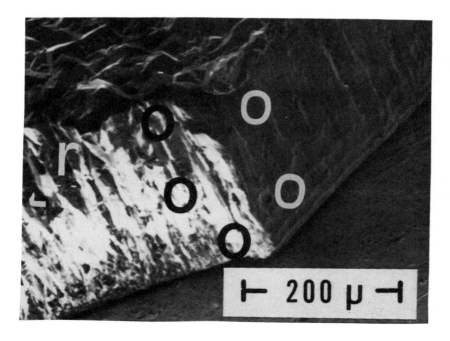

Fig. 5. Transverse optical micrograph of a G.E. diamond-like film, showing distinct layers on the right (laser cut) edge, but a uniform structure for the left (cleaved) edge.

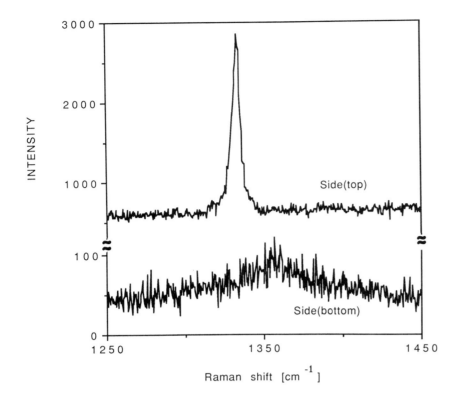

Fig. 6. Raman scattering spectra for a G.E. diamond-like film. The upper curve, taken at the top of the laser-cut edge (see Fig. 5), shows a strong 1331 cm^{-1} line, while the lower spectrum, taken at the lower part of that edge, shows a much broader line at about 1360 cm^{-1}.

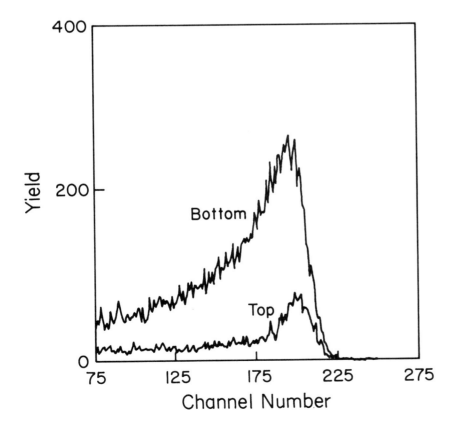

Fig. 7. Elastic recoil spectra recorded for the top and bottom surfaces of the G.E. sample of Figs. 5 and 6.

4. DISCUSSION AND CONCLUSIONS

The present results have demonstrated that optical microscopy, elastic recoil detection spectroscopy and Raman scattering are a useful combination of techniques to characterize diamond-like films. Obvious surface and bulk features visible by microscopy were related to the H content by ERD and to the presence of true diamond material by Raman scattering.

5. ACKNOWLEDGMENTS

The authors thank R.A. Rudder (RTI), S.H. Yokota (Crystallume Corp.) and T.R. Anthony (G.E. Co.) for supplying the diamond and diamond-like films.

This work is supported by the Office of Naval Research under contract N00014-87-K-0243.

6. REFERENCES

1. Y. Saito, S. Matsuda and S. Nogita, J. Mater. Sci. Letters 5, 565 (1986).
2. M. Tsuda, M. Nakajima and S. Oikawa, J. Amer. Chem. Soc. 108, 5780 (1986).
3. Proc. 8th Int. Symp. on Plasma Chemistry, Tokyo 1987.
4. K. Kurihara, K. Sasaki, M. Kawarada and N. Koshino, Appl. Phys. Lett. 52,437 (1988).
5. D.J. Vitkavage, R.A. Rudder, G.G. Fountain and R.J. Markunas, J. Vac Sci. Technol. A6, 1812 (1988).
6. A. Turos and O. Meyer, Nucl. Instr. Meth. B4, 92 (1984).
7. S.A. Solin and A.K. Ramdas, Phys. Rev. B1, 1687 (1970).
8. B.S. Elman, M.S. Dresselhaus, G. Dresselhaus, E.W. Maby, and H. Mazurek, Phys. Rev. B25, 4142 (1982).

SESSION 5

Diamond Applications

Chair
Sidney Singer
Los Alamos National Laboratories

Invited Paper

ARTIFACT DIAMOND
Its Allure and Significance

Max N. Yoder

Electronics Division
Office of Naval Research
Arlington,Virginia 22217

ABSTRACT

While the preponderance of the mechanical, optical, and electronic properties of natural diamond have been known for over a decade, only recently has artifact diamond in technologically useful form factors become an exciting possibility. The advent of sacrificial, lattice matched crystalline substrates provides the basis not only for semiconducting applications of diamond, but for optical mirrors, lenses, and windows as well.

As a semiconductor, diamond has the highest resistivity, the highest saturated electron velocity, the highest thermal conductivity, the lowest dielectric constant, the highest dielectric strength, the greatest hardness, the largest bandgap and the smallest lattice constant of any material. It also has electron and hole mobilities greater than those of silicon. Its figure of merit as a microwave power amplifier is unexcelled and exceeds that of silicon by a multiplier of 8200. For integrated circuit potential, its thermal conductivity, saturated velocity, and dielectric constant also place it in the premier position (32 times that of silicon, 46 times that of GaAs). Although not verified, its radiation hardness should also be unmatched.

Aside from its brilliant sparkle as a gemstone, there has been little use of diamond in the field of optics. Processing of the diamond surface now appears to be as simple as that of any other material -- albeit with different techniques. In fact, it may be possible to etch diamond far more controllably (at economically viable rates) than any other material as the product of the etch is gaseous and the etched trough is self-cleaning. Other properties of diamond make it an ideal optical material. Among them are its unmatched thermal conductivity, its extremely low absorption loss above 228 nanometers, and unmatched Young's modulus, Poisson's ratio, tensile strength, hardness, thermal shock, and modulus of elasticity. If the recently-found mechanisms by which erbium impurities in III-V junctions can be made to "lase" are indeed applicable to indirect gap semiconductors, then even efficient diamond lasers at attractive portions of the spectrum become a distinct possibility.

1. BACKGROUND

Although diamond has been know for centuries, it was the advent of X-rays that permitted its crystalline structure to become known in the second decade of the present century. The father and son Bragg duo

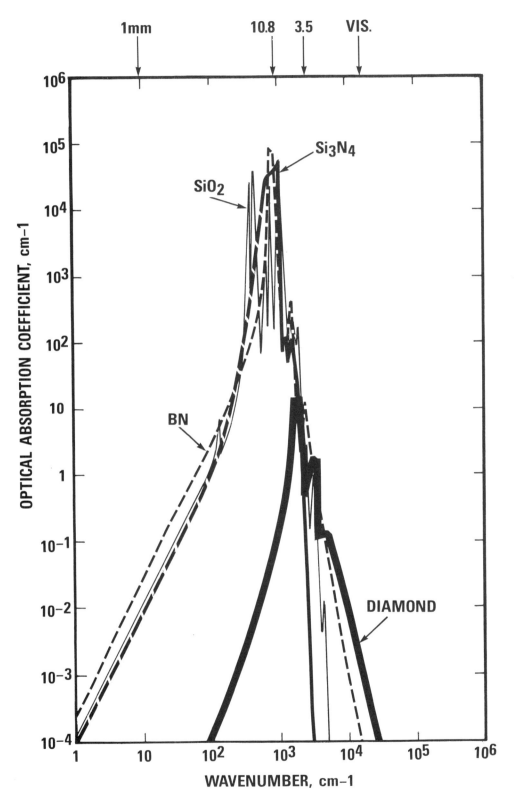

Figure 1.   Optical absorption in diamond and other materials.

used X-ray diffraction to show that diamond had a cubic structure and a lattice constant of 3.56 Angstroms. With tetrahedral bonding, the nearest neighbor distance is 1.53 Angstroms and at the centerpoint the electron density is approximately 1.59 electrons per square Angstrom! Although the covalently bonded diamond crystal exhibits a very open lattice, the presence of an interstitial impurity atom, a misplaced carbon atom, or a vacancy leads to a considerable change in the electron density and to lattice distortion.

2. CHARACTERIZATION

With an electronic bandgap of 5.45 electron volts, diamond should exhibit no optical absorption within the visible spectrum as its absorption edge should be 2250 Angstroms. Although many water clear gemstones are found in nature, there are also those of yellow, blue, green, pink, mauve, and black. Clearly nature has placed electronic absorption levels within the bandgap. It was these in-gap absorption levels that led diamonds to be broadly classified as type 1 or type 2. If the specimens demonstrated a sharp increase in absorption at wavelengths shorter than 2250 Angstroms and there was negligible absorption at wavelengths longer than 3000 Angstroms, they were classified as being type 2. Type 2 diamonds exhibit very little visible fluorescence under moderate ultra violet or ionizing illumination. Under intense ionizing radiation, even type 2 diamonds will exhibit luminescence in the blue region. Although ultraviolet emission has been seen, it is very rare. In type 2 diamonds, the response to radiation near to but below the bandgap is generally that of its response to radiation of much longer wavelengths. In this case the dielectric constant throughout the visible spectrum is equal to the static dielectric constant (5.76) or the square of the index of refraction. Type 2 diamonds are further subdivided as A or B with the A group having a very sharp and temperature independent absorption cut off (they are P-type semiconductors) and the B group having a broader and temperature dependant absorption cut off. Type 1 diamonds exhibit significant absorption at wavelengths of 3000 Angstroms or longer. Excepting for their tribological use as abrasives or cutting tools, they have little commercial usefulness. Although diamond is monatomic and non-polar, even the best of natural diamonds exhibit some infrared absorption centered at 7.8 micrometers. Even then, its absorption is approximately 4 orders of magnitude less than that of competing materials as shown in figure 1.[1]

3. MIRRORS, LENSES, AND COATINGS

Figure 1 compares the optical absorption of diamond with several competing materials for optical lenses and windows. Although not shown in the figure, diamond also exhibits approximately four orders of

---

[1]. Data for figure 1 supplied by R. Schwartz of the Naval Weapons Center

magnitude less peak absorption than do magnesium difluoride, yttrium oxide, and zinc selenide. For use with high intensity optical beams such as generated by free electron lasers, diamond has another unmatched attribute; its thermal conductivity is unexcelled. Thus even if appreciable energy is absorbed, the absorbed energy can be readily dissipated to the edge of the diamond. As such, optical beams exhibiting power densities greater than 10^6 Watts/cm^2 can be transmitted by diamond windows.[2]

Liquid etches, even at high temperatures, are well known to be ineffective in etching diamond. While atomic oxygen is known to attack diamond, it is not an efficient etch. Using a combination of energetic ions and a physisorbed, oxygen-containing gas, Geis has shown that the diamond surface can be etched as easily as that of any other semiconductor.[3] Not only has he been able to create etch troughs ten times deeper than the remaining diamond pillars are wide, the residual effluent is gaseous and the trough bottoms are self-cleaning! The feasibility of a diamond Fresnel lens is thus established. Using special high power laser processing techniques, Geis has also demonstrated that diamond windows can be fabricated as optical polarizers giving extinction ratios in excess of 15:1 at temperatures as high as 1400 Celsius.[4]

While it is difficult to conceive of the use of natural diamonds to protect optical mirrors, lenses, and windows made of other materials, the advent of synthetic diamond engenders the optical engineer with better approaches to systems design. Not only can the underlying optical material be protected against microparticle abrasion by virtue of the unmatched hardness of diamond (9000 kG/mm^2), but its extremely high surface energy (as high as 9400 ergs/cm^2) protects it from many (including biological) adherents -- virtually nothing (excepting hydrogen) wets diamond. To improve the threshold of high power optical damage to silicon windows, a consortium of investigators from Crystallume, NASA Langley Research Center, and Old Dominion University investigated synthetic diamond overcoatings.[5] They used the thermal stress parameter to motivate their work. It is proportional to tensile fracture strength, thermal conductivity, thermal coefficient of expansion, and elastic modulus -- in each of which the properties of

2. J. R. Seitz, "Laser systems with diamond optical elements", U. S. Patent 3,895,313 15 July 1975
3. M. W. Geis, presented at the 1986 S.D.I.O./IST - ONR Diamond Technology Initiative Seminar, Durham, N.C. 17 July 1986 (unpublished).
4. M. W. Geis, "Electrical and optical properties of diamond interfaces", Proceedings, Third Annual SDIO/IST-ONR Diamond Technology Symposium, Crystal City, VA 12-14 July (1988), Paper W15
5. S. Albin, A. Cropper, L. Watkins, C. Byvik, A. Buoncristiani, K. Ravi, and S. Yokota, "Diamond films for laser optics", Paper Th4 Book of Abstracts, SDIO/IST-ONR Diamond Technology Symposium, 12-14 July 1988, Crystal City, VA

diamond are unexceeded. The measured threshold of single pulse 1.06 micrometer illuminating damage ranged from a low of 5.97 J/cm² to a high of 12.43 J/cm². Comparisons of the thermal stress parameter of diamond with that of other optical materials is shown in figure 2.

4. ELECTRONICALLY ACTIVE DIAMOND

Virtually all of the electronics properties of diamond are unexcelled excepting that of electron mobility. Even then, its mobility exceeds that of silicon. With very high charge carrier mobilities and velocities, diamond is expected to exhibit extremely fast electronic devices capable of operating with very large voltage excursions, at very high temperatures, and with high immunity to radiation damage.[6] Unfortunately, diamond is an indirect bandgap semiconductor and it is well known that injection lasers fabricated with indirect bandgap materials have not been successful. Diamond, however, affords a new approach. With its extremely large forbidden bandgap, there exists the possibility that many elements of the periodic table may have optical transition energies therein, as blue emission and even externally pumped lasing is well know in diamond.[7] In recent years, charge carrier pumping of impurities in more common semiconductors has produced emission and even single mode lasing.[8,9] Using similar techniques it is feasible that diamond can be internally pumped by forward biasing a diamond junction or, more plausibly, a heterojunction of boron nitride on diamond. Since diamond is a column IV semiconductor, a valence 3 rare earth impurity could serve as the acceptor material immediately inside the P side of such a junction and possess a very large capture cross section for "pump" electrons. Figure 3a illustrates such a rare earth atoms shown as A3+ before forward bias while figure 3b illustrates a possible electron and optical transition after application of the forward bias. In this illustration 4f transitions of rare earth atoms are shown, but 2d transitions of the transition metals impurity atoms may also be considered. Figure 4 illustrates transition levels from impurity atoms having energy levels just within the conduction and valence bands. While less probable than those transitions shown in figure 3, they are possible. Finally, optical transitions from impurity levels just within the conduction band to energy levels within the same ion, but

6. M. N. Yoder, "Synthetic diamond, its properties and synthesis", Mat. Res. Soc. Symp. Proc. 97, 315-326 (1987)
7. S. Rand and L. DeShazer, "Visible color-centered laser in diamond", Optics Lett. 10(10) 481-483 (1985)
8. P. Klein, J. Furneaux, and R. Henry, "Time-dependant photoluminescence of InP:Fe", Phys. Rev. B 29(4), 1947-1961 (1984)
9. W. Tsang and R. Logan, "Observation of enhanced single longitudinal mode operation in 1.5 micron GaInAsP erbium-doped semiconductor injection lasers", Appl. Phys. Lett. 49(25) 1686-1688 (1986).

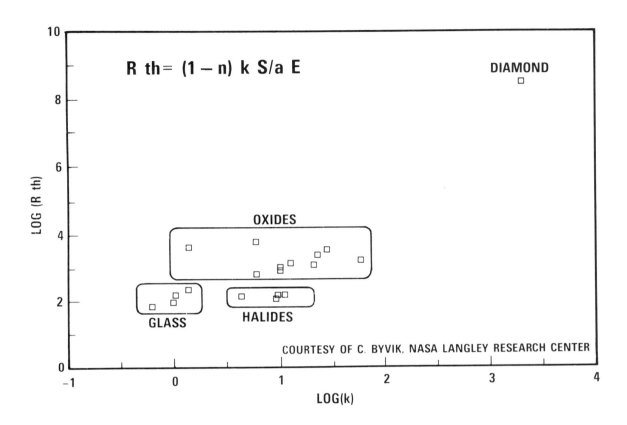

Figure 2.   Thermal stress parameters of diamond and other materials.

within the diamond bandgap can not be ignored. Thus it is conceivable that a diamond injection laser could be engineered to emit at virtually any wavelength from the infrared to the ultra violet.

5.   SUMMARY

The combination of high tensile strength, unexcelled thermal conductivity, low optical absorption, very large bandgap, appropriate Young's modulus and Poisson's ratio, and recent advances in its synthesis render it an ideal candidate for many optical applications.

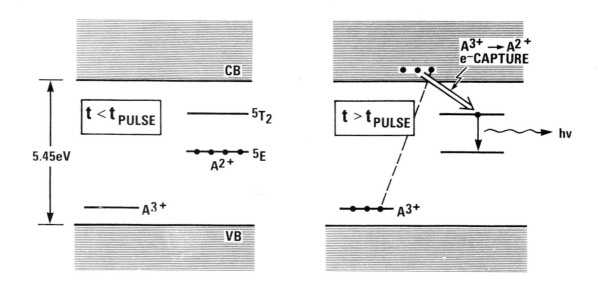

Figure 3.   Intra ion energy level transitions in diamond.
(a) before forward bias, (b) after bias pulse

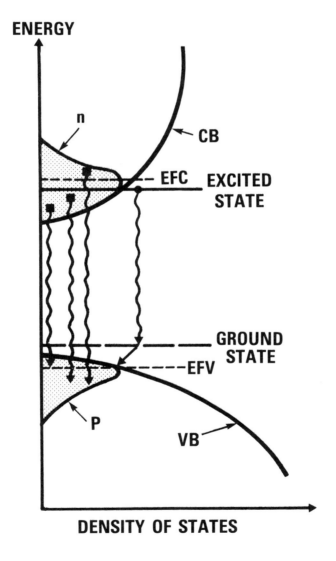

Figure 4. Bandgap magnitude intra ion energy transitions

The separation of natural from synthetic diamonds on the
basis of their optical and microscopic properties

Emmanuel Fritsch, James E. Shigley, John I. Koivula

Research Department, Gemological Institute of America
1660 Stewart Street, Santa Monica, California 90404

ABSTRACT

The commercial availability of gem-quality synthetic diamonds, and the expanding research
on various methods of diamond synthesis, have made it necessary for the gem diamond trade to
find simple ways to identify synthetic diamonds. Using both magnification and various types
of spectroscopy and luminescence techniques, a number of properties that are characteristic
of synthetic diamonds have been found.

1. INTRODUCTION

In 1985, synthetic diamonds of a size and quality suitable to be cut into attractive gemstones became commercially available for the first time. This material has created a new
concern for the gem diamond trade, as one needs to reliably identify the synthetic origin of
a stone, even after it has been faceted, in order to avoid selling a synthetic gem diamond
as a natural one.

An additional cause of anxiety is the possible application of the low-pressure synthesis
of diamond thin films to produce a coating on a faceted gemstone that could change the gemological properties of the material in some as yet unanticipated ways (such as improving the
color).

Non-destructive optical studies of synthetic diamonds grown at either low or high pressures by different manufacturers were undertaken using microscopic and spectroscopic
methods. Results of our examination demonstrate that these materials, whether in the form
of crystals or thin films, have distinctive properties that distinguish them from natural
diamonds.

2. DIAMOND CRYSTALS MANUFACTURED BY A HIGH-PRESSURE, HIGH-TEMPERATURE PROCESS

We examined synthetic diamonds produced by the General Electric Company (GE)[1] (yellow,
blue, and colorless), Sumitomo Electric Industries (SE)[2](yellow), and the De Beers Diamond
Research Laboratory (DB)[3](brownish yellow, yellow, and greenish yellow). The synthetic diamonds fall into two categories - the yellow material is nitrogen-containing type Ib, while
the blue and colorless materials are type IIb diamond containing a small amount of boron.

2.1 Flux inclusions

When examined with an optical microscope, the crystals of synthetic diamond were found
to contain small (approximately 0.1 to 1 mm) rod-like black inclusions that have a metallic
appearance. These black inclusions are the most obvious feature indicative of laboratory
growth, since inclusions of this kind are not found in natural diamonds. The inclusions are
mostly iron (GE) or an iron-nickel alloy (SE and DB), as determined by electron microprobe
analysis. They represent solidified remnants of the molten metal flux material from which
the synthetic diamond crystals are grown. Their shape, appearance, and crystallographic
orientation within the synthetic diamond host help in distinguishing them from the black
inclusions of spinel or sulfide minerals that occur in some natural diamonds. In addition,
tiny white inclusions, sometimes arranged in "broom"-shaped (GE) or geometric patterns (DB),
are also common in these synthetic diamonds. In one instance, these tiny white inclusions
proved to be small particles of iron (DB). Although small, "pinpoint" inclusions also occur
in many natural diamonds, their appearance is quite different from the white inclusions described here.

2.2 Sectorial growth

The high-temperature, high-pressure growth process creates diamond crystals with a
regular arrangement of internal growth sectors. Small cubic and trapezohedral growth sectors that are observed in the synthetic crystals have never been seen in natural diamonds,
in addition to the octahedral and dodecahedral growth sectors that are present in the morphology of both natural and synthetic diamond crystals. The boundaries of these internal
growth sectors are often marked by surfaces of higher index of refraction. These surfaces
can often be visually observed with the microscope, and are then referred to by the term

"graining" by gemologists. These internal growth sectors are particularly noticeable in the dark yellow synthetic diamonds. They give rise to visual patterns of "graining" never before seen in faceted natural yellow diamonds. By looking through the lower portion, or pavilion, of a faceted dark yellow synthetic diamond, one can observe a "graining" pattern reminiscent of an "hourglass" design. A "graining" pattern of intersecting lines in an octagonal shape is observed using low-incidence angle reflected light on the large polished facet, or table, on the top side of a faceted dark yellow synthetic diamond. This surface "graining" pattern arises from the greater hardness (diamond harder than diamond!) at growth sector junctions or within certain sectors themselves. These "graining" patterns are less obvious as the nitrogen content of the diamond decreases, suggesting that they are related to nitrogen impurities, and more importantly, that they might not exist in colorless type IIb (or IIa) synthetic diamonds which contain little or no nitrogen.

Color zoning is another characteristic feature in crystals of both yellow and blue synthetic diamonds, again related to the sectorial nature of crystal growth. Different color zones are separated by often sharp and geometrically-arranged boundaries that generally follow the "graining". The color zoning patterns are quite different from what is observed in natural diamonds. However, it might be difficult to observe after the synthetic diamond has been faceted.

The internal growth sectors are also clearly seen using either short-wave ultraviolet luminescence or cathodoluminescence. In the yellow synthetic diamonds, cubic sectors display a strong yellow to yellowish-green fluorescence while adjacent octahedral sectors remain inert. Cathodoluminescence photographs provided by De Beers of their synthetic diamonds demonstrate that the trapezohedral growth sectors luminesce only weakly. Similar luminescence patterns have not been observed in natural diamonds.

When examined with a microscope between crossed polarizers, natural diamonds often exhibit anomalous birefringence (improperly called "strain" by gemologists) usually in planes parallel to the octahedral crystal faces. In contrast, the synthetic diamonds show quite different (and often weaker) anomalous birefringence patterns related to the internal growth sectors.

2.3 Optical properties

2.3.1 Ultraviolet luminescence.
One of the most time-efficient ways to separate natural from synthetic diamonds is to check their luminescence to both long-wave (LW) and short-wave (SW) ultraviolet radiation. Synthetic yellow diamonds are inert to LW, but have various responses to SW (GE: inert; SE: greenish yellow; DB: inert or yellow, depending upon the body color). Synthetic blue diamonds are inert to LW and either bluish white or yellow (GE), or greenish yellow (SE) to SW; in both instances they also have a persistent phosphorescence. Synthetic colorless diamonds are inert to LW and are yellow (GE) or greenish yellow (SE) to SW, and again display a persistent yellow phosphorescence. In contrast, natural diamonds that are fluorescent have a stronger response to LW than to SW, and have varied colors of luminescence such as blue, yellow, or green.

2.3.2 Ultraviolet/visible absorption.
No sharp absorption bands are observed in the visible-range spectra of the synthetic diamonds. In the yellow synthetic diamonds, there is only a rapid increase in absorption toward the ultraviolet beginning at 550 nm. This contrasts with most (but not all) dark yellow natural diamonds whose spectra display a series of one or more sharp bands (the "Cape spectrum" lines) at 415, 435, 452, and 478 nm arising from the N3 and N2 centers. In the blue synthetic diamonds, there is an increasing absorption toward the infrared, as is the case for natural blue diamonds.

2.3.3 Infrared absorption.
All synthetic yellow diamonds are relatively pure type Ib, as shown by their absorption in the range from 1000 to 1500 wavenumbers. The intensity of the absorption due to isolated nitrogen is similar for all samples of synthetic diamond of a given color and manufacturer. In contrast, type Ib natural yellow diamonds are extremely rare; they usually have a noticeable type Ia component in their infrared spectra. In addition, the total amount of nitrogen in these natural yellow diamonds is highly variable. Blue and colorless synthetic diamonds are relatively pure type IIb.

2.4 Magnetic properties

After proper cleaning, natural diamonds are less magnetic than the synthetic GE diamonds (with an exception of a near-colorless faceted sample). In synthetic diamonds, this is attributed to a high concentration of magnetic flux inclusions, some of which are visible under the microscope. However, preliminary experiments on the SE faceted yellow samples indicate that they have a very low magnetism, and therefore, probably a very low amount of flux inclusions.

3. DIAMOND THIN FILMS MANUFACTURED AT LOW PRESSURE (CVD-TYPE DEPOSITION)

Currently a great deal of research is being carried out with the goal of producing a thin diamond film on a substrate. While this thin film technology has not yet been directed toward the coating of gem materials, the likelihood of this application in the future seems high.

Polycrystalline diamond thin films deposited at low pressure are easily recognizable due to their granular texture. If the individual grains are too small to produce a haziness (that results from light scattering), Nomarski differential interference contrast helps to visualize them. In addition, the smoothness of the polished surface of a faceted gemstone is much greater than that of a surface coated with a thin film.

4. CONCLUSION

Yellow and blue synthetic diamonds are easy to separate from their natural counterparts on the basis of the existence of cubic growth sectors. In yellow synthetic diamonds, sectors or sector boundaries are highlighted by many nitrogen-related features (such as color zoning and "graining") that makes them easy to identify. However, the separation of near-colorless synthetic diamonds from similar natural diamonds might be more challenging as the nitrogen-related features become less obvious with lower nitrogen content. At this time, all near-colorless synthetic diamonds we have examined have a weak type IIb character.

5. ACKNOWLEDGEMENTS

The authors wish to thank the research staff of both Sumitomo Electric Industries and the De Beers Diamond Research Laboratory for the loan of samples of synthetic diamond and for information on the diamond synthesis process. Dr. Kurt Nassau and Mr. John S. White (of the National Museum of Natural History, Smithsonian Institute) both provided us samples of General Electric synthetic diamond. Ms. Laurie Conner and Dr. Michael Pinneo of Crystallume loaned us a sample of a diamond thin film and have been an invaluable source of information. Dr. George Rossman (California Institute of Technology) arranged for the magnetism studies of the synthetic diamonds. The research and laboratory staff of the Gemological Institute of America provided numerous constructive comments.

6. REFERENCES

1. J.I. Koivula and C.W. Fryer, "Identifying gem-quality synthetic diamonds: an update", Gems and Gemology, 20(3), 146-158 (1984).
2. J.E. Shigley, E. Fritsch, C.M. Stockton, J.I. Koivula, C.W. Fryer, and R.E. Kane, "The gemological properties of the Sumitomo gem-quality synthetic yellow diamonds", Gems and Gemology, 22(4), 192-208 (1986).
3. J.E. Shigley, E. Fritsch, C.M. Stockton, J.I. Koivula, C.W. Fryer, R.E. Kane, D.R. Hargett, and C.W. Welch, "The gemological properties of the De Beers gem-quality synthetic diamonds", Gems and Gemology, 23(4), 187-206 (1987).

Some Properties of Synthetic Single Crystal and Thin Film Diamonds

Shuji Yazu, Shuichi Sato, Naoji Fujimori

Sumitomo Electric Industries Ltd., Itami Research Laboratories
1-1-1, Koyakita, Itami, Hyogo 664, Japan

ABSTRACT

Large synthetic diamond single crystals, in sizes up to 1.4 ct, are produced on a commercial basis for some industrial application fields by Sumitomo Electric. The crystals are yellow colored type Ib stones which contain lower amounts of nitrogen (up to about 100 ppm) dispersed through the crystal structure in the form of singly substituting atoms. The impurity controlled type Ib crystals have the highest thermal conductivity which is equivalent to that of pure type IIa crystals. Optical and thermal properties of diamond crystals are strongly affected by dispersed impurities. We studied the kinds of dispersed impurities and amounts of those impurity atoms in our synthesized crystals by SIMS. A relation of the thermal conductivities and the nitrogen concentrations of the crystals was examined. The state of nitrogen impurity in the crystals could be transformed by electron irradiation and subsequent high temperature annealing. The reaction rates for the transformation Ib nitrogen to type IaA aggregates and differences in crystal growth sectors have been studied.

Vapor phase deposited diamond films are hopeful candidates for optical application of diamond. Preliminary spectroscopic analysis has been done for the free standing polycrystalline films.

1. INTRODUCTION

Progress in high pressure technology yielded a mass production system of high quality single crystal diamonds. Sumitomo Electric started production of yellow single crystal diamonds in 1985. Those crystals are applied as heat sinks for semiconductor devices such as laser diodes and IMPATT diodes, wire drawing dies, tool blanks for precision diamond turning and others. (See Figure 1.) Although specific optical application has not emerged, it would be useful to recognize thermal and optical properties of the synthetic crystals compared with natural stones for finding out suitable application.

High pressure synthesized diamonds are classified type Ib, IIa and IIb. (See Table 1.) In this study, we used type Ib stones which were picked up from the mass production line and type IIa and Type IIb crystals which were grown in our laboratory. Optical properties of diamond are mainly affected by the state of atomically dispersed impurities. In the following sections, we describe about the amounts of dispersed impurities in high pressure synthesized crystals, thermal and optical properties particularly related the state of nitrogen impurity.

2. DISPERSED IMPURITIES IN SYNTHETIC CRYSTALS

Concerning about the quality of synthetic diamond crystals, a major problem is metallic inclusions. The amount of the inclusions can be minimized by controlling precisely the pressures and temperatures during synthesis. In this study, we don't care about the metallic inclusions. To clarify the atomicaly dispersed impurities in our synthetic crystals, we used SIMS (Secondary Ion Mass Spectroscopy).

The preparation conditions of sample type Ib and type IIa crystals are listed in Table 2. The samples were grown by "temperature gradient method" at high pressures using molten metal solvents.[1] SIMS (Cameca IMS3f) analysis suggested that main impurities in the crystals were solvent metal atoms except for nitrogen in the type Ib crystal. For quantitative measurements of Ni, Fe and Al atoms in the crystals, fixed amounts (1×10^5 atoms/cm^2) of those atoms were ion implanted in the surface of the crystals and the concentrations of the impurities were determined comparing with the depth profiles of implanted atoms. The results of the quantitative measurements are listed in Table 3. It should be noted that Ni and Al atoms were detected in the amounts of more than 10 ppm in type Ib and type IIa crystals respectively, but Fe concentrations are around 1ppm in the both types of crystals. Collins suggested that diamonds grown using a nickel alloy solvent showed sharp absorption at the Raman frequency (1332 cm^{-1}) and nickel could be optically active in diamonds.[2] In our case, such a strong absorption was not observed. Aluminum was originally thought to be the acceptor as well as boron. Chrenko showed that aluminum was almost certainly present as metallic inclusions in synthetic diamond grown aluminum included metallic solvents.[3] Our type IIa sample could not be detected any inclusions under optical microscope examinations and the secondary ion mass spectra were consistent at the several points of the examined crystal, and so we think there would be a possibility aluminum to be dispersed in the crystal structures. More detailed studies have to be done for understanding the state of the impurity atoms in the crystal structures. The total impurity concentrations of the synthesized stones are rather low compared with those of pure natural type IIa crystals. [2]

3. THERMAL CONDUCTIVITY OF SYNTHETIC CRYSTALS

Diamond conducts heat better than any other material at room temperature. Burgemeister studied a relation of the thermal conductivity of natural diamond and its nitrogen concentration. He suggested that the thermal conductivity of natural stones was strongly correlated with the total amounts of nitrogen, which were estimated from infrared absorption coefficients, in the crystals. [4] Pure natural type IIa diamond which contains less than 1pp nitrogen has a thermal conductivity of about 20 Watts/°C·cm at room temperature. Type Ia diamond, most of natural gem and industrial stones classified to this type Ia, has a much lower value of a thermal conductivity, because of the higher nitrogen concentration. [5]

Slack measured the thermal conductivity of synthesized single crystals. [5] His measurements gave slightly higher thermal conductivities for type IIa (less than 1 ppm nitrogen) and type Ib (about 50 ppm nitrogen) synthetic crystals than the reported values of pure type IIa natural stones at room temperature.

We conducted the measurements of the thermal conductivity of synthesized crystals which contained different amount of dispersed nitrogen. The equipment which we used was specially designed one to measure the small diamond bars or cubes by determining temperature gradients with an InSb radiation detector under the steady state heat flow. The principle of this type of equipment was described by Burgemeister. [5] Typical dimensions of sample crystals were 2x2x3 or 2x2x4 mm bars. The nitrogen concentration in the synthetic crystals was estimated by infrared absorption spectra at 1130 cm^{-1} using the correlation proposed by Chrenko et al. [6]

Figure 2. shows the thermal conductivities of synthesized type Ib, natural IIa and Ia crystals as a function of temperature. This measurement gave high thermal conductivities for type Ib synthetic crystals equivalent to the value of pure natural type IIa diamond in this temperature range. The result of the measurements of synthetic stones, which contained dispersed nitrogen in the rage of 18 - 126 ppm, at 45 °C is shown in FIgure 3. In this experiment, the thermal conductivities of type Ib crystals which have lower than about 85 ppm of nitrogen were constant and comparable to that of natural type IIa crystal. It is reported that the thermal conductivities of natural type Ia stones drop lineally with increasing nitrogen concentrations. [4] This discrepancy suggests that aggregated state of nitrogen in natural crystals has strong influence to phonon scattering comparing with dispersed state of nitrogen in synthesized crystals.

4. TRANSFORMATION OF OF THE STATE OF NITROGEN IN SYNTHETIC CRYSTALS

Synthetic type Ib (contain isolated substitutional nitrogen) crystals can be transformed to type IaA (contain nitrogen in the A aggregate form) or mixed type by high temperature-high pressure annealing. [6] Collins described that the aggregation process could be greatly enhanced by the presence of vacancies introduced by electron irradiation. [7] The study in this connection will provide information about the way to control the optical properties of synthetic diamond.

Samples which we used were plates sliced from about 1 ct synthesized stones and contained dispersed nitrogen in the range of 40 - 100 ppm. High temperature-high pressure treatments were conducted using high pressure apparatus under 5.0 GPa and the temperature range of 1700 - 2100 °C. Some samples were irradiated with 10^{18} 2 MeV electrons/cm^2 and heat treated at 5.5 GPa and 1750°C for 50 hours. We measured the change of dispersed nitrogen content using ESR (Electron Spin Resonance) instead of absorption spectra because of sensitivity. The integrated intensity of the center line of the dispersed nitrogen hyperfine triplet in ESR spectra was a measure of dispersed nitrogen content.

Infrared absorption spectra for a synthetic type Ib crystal, as grown and heat treated under pressure at 1750°C for 50 hours after irradiation, are shown in Figure 4. Dispersed nitrogen in type Ib diamond and IaA aggregated nitrogen induce the infrared absorption at 1130 and 1280 cm-1 respectively. In Figure 4., the partial conversion of the type Ib to type Ia crystal is shown. Chrenko and others suggested that the transformation was controlled by second order diffusion kinetics and the rate constant k was expressed as the following equation.

$$kt = 1/c - 1/c_o$$

in which t is the time; c and c_o are the initial concentration and the remaining concentration after heat treatment of type Ib nitrogen respectively. The transformation rate constant k depends on temperatures. The results of our experiments are summarized in Figure 5. For unirradiated samples, the rate constant at 1700°C is low and no detectable reduction of the 1130 cm^{-1} band was observed, while the rate constant at the same temperature for irradiated samples was 1.4×10^{-5}. The rate constant for unirradiated crystals at 2100°C was 1.1×10^{-5}. As shown in Figure 5. electron irradiation enhanced the transformation at temperatures below 2000°C, but the effect was not identified at the higher temperatures. Probably, the defects induced by electron irradiation could be annealed rapidly at the higher temperatures.

In the course of the above described experiments, we found that the nitrogen transformation rate was clearly different with growth sectors in a crystal. Figure 6. shows a typical growth sectors in a synthesized type Ib diamond grown by temperature gradient method. Figure 7. is the UV-Visible range absorption spectra of the tested samples. In the figure, the spectra of as grown crystals and the spectra of (100) and (111) sectors are shown. In (100) sectors, the transformation rate of dispersed state of nitrogen to aggregated form was lower than in the case of (111) sectors, while the formation of N3 center was dominant. This experiments provide some information for understanding the structure of N3 center and the mechanism of nitrogen incorporation in synthetic type Ib crystals.

5. OPTICAL TRANSMISSION OF POLYCRYSTALLINE DIAMOND FILMS

Low pressure diamond synthesis technology attracts grate interest of many researchers. In Japan, much efforts are concentrated on the development of higher growth rate processes and large area deposition processes. It seems to be important to get pure thick polycrystalline diamond films at a reasonable growth rate. We describe the very preliminary test results which were conducted in this direction.

Diamond film samples were prepared on Si and Mo substrate from a gaseous mixture of hydrogen and methane under microwave glow discharge conditions. The thickness of the diamond films were 10 - 60 microns depending on the deposition rate and duration. After dissolving the substrates, UV-Visible range transmission were measured using a spectroradiometer (LI-COR LI-18000) with an integrating sphere for gathering transmitted light. Typical transmission spectra of black and white diamond films are shown in Figure 8. The white translucent films had the absorption edge corresponding with the intrinsic diamond absorption edge 225 nm. THe black films did not transmit shorter wave length light than 500 nm and exhibited a broad absorption band in the 1000 nm range. We examined the surface morphology of the film samples with SEM and found that the translucent films consisted of uniform large diamond grains (about 10 micron or more), while the grains of the black films were leaflike as shown in Figure 9.

6. CONCLUSIONS

High pressure synthesized diamond crystals have the high thermal conductivity equivalent to the value of pure natural type IIa stones and the optical properties can be controlled to some extent by irradiation and heat treatment modifying the state of nitrogen impurity. Progress in low pressure diamond synthesis technology will bring new optical application of synthetic diamond.

7. REFERENCES

1. H.M. Strong and R.H. Wentorf, Jr., Naturwissenscaften, 59.Jg., Heft 1, 1-7 (1972).
2. A.T. Collins and P.M. Spear, J.Phys. D: Appl. Phys., 15, L183-187 (1982).
3. R.M. Chrenko, Phys. Rev. B 7, 4560-7 (1973).
4. E.A. Burgemeister, Physica, 93B, 165-179 (1978).
5. E.A. Burgemeister, Ind. Diamond Rev., July, 242-244 (1975)
6. R.M. Chrenko, R.E. Tuft and H.M. Strong, Nature 270, 141-144 (Nov. 1977).
7. A.T. Collins, J. Phys. C: Solid St. Phys., 13, 2641-50 (1980).

Table 1. Classification and properties of diamonds.

TYPE	I			II	
	Ia	Ib		IIa	IIb
YIELD RATIO OF NATURAL DIAMOND	98 %	0.1 %		1~2 %	
IMPURITY (ppm) — N	2×10^3 Platelet	$10^2 \sim 10^3$ Dispersed	$1 \sim 10^2$	~1	
IMPURITY (ppm) — OTHERS		Catalyst metals $10^3 \sim 10^5$	~10		B ~10^2
COLOR	White-Yellow	Green-Brown	Yellow	White	Blue
ELECTRIC RESISTIVITY (Ω·cm)	$10^4 \sim 10^{16}$	$10^4 \sim 10^{16}$	10^{16}	10^{16}	$10 \sim 10^4$
THERMAL CONDUCTIVITY (W/m·K)	900	900	2000	2000	
SYNTHETIC DIAMOND		Powder	Crystal		

Table 2. Preparation conditions of synthetic crystals.

Sample No.	Type	Solvent Composition (wt%)	Crystal growth condition Temperature (°C)	Pressure (GPa)	Nitrogen concentration (ppm)
1	Ib	97Fe-3Al	1400°C	5.5	< 1
2	IIa	50Fe-50Ni	1350°C	5.5	32

Table 3. Results of quantitative impurity analysis.

Sample No	Concentration of atoms (atoms/cm³)		
	Al	Fe	Ni
1	3.19×10^{18}	2.11×10^{17}	2.31×10^{17}
2	1.60×10^{17}	1.13×10^{17}	2.08×10^{18}

Figure 1. Synthetic diamond single crystals and heat sinks.

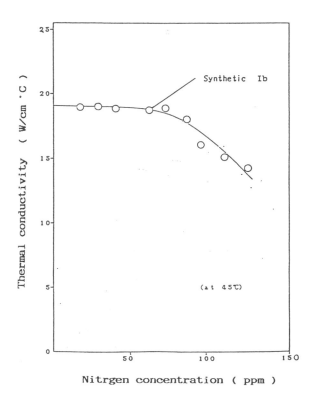

Figure 2. The thermal conductivities of synthetic and natural diamonds as a function of temperature.

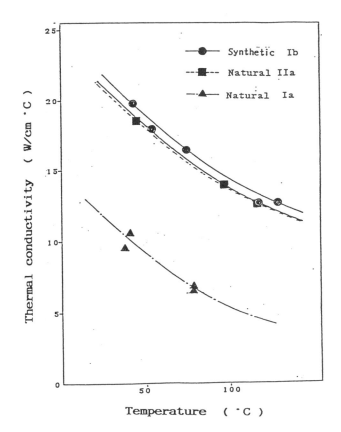

Figure 3. The thermal conductivities of synthetic crystals as a function of the nitrogen concentrations.

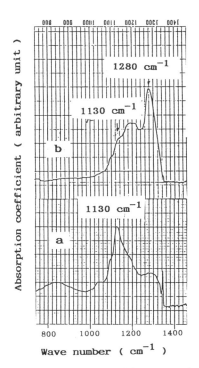

Figure 4. Infrared absorption spectra for a, a type Ib synthetic diamond as grown; b, the same crystal heat treated under pressure at 1750°C for 50 hours after irradiation.

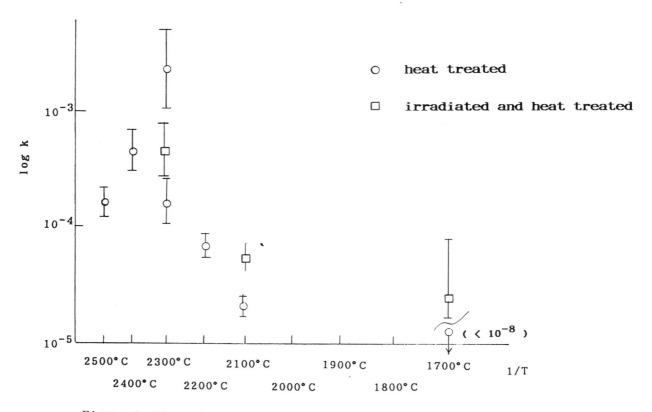

Figure 5. The rate constant k at various annealing temperatures.

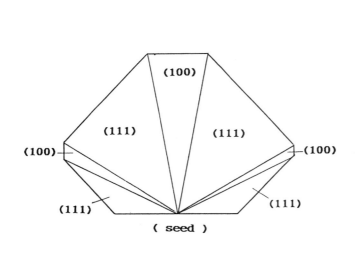

Figure 6. A typical growth sectors in a synthesized Ib crystal.

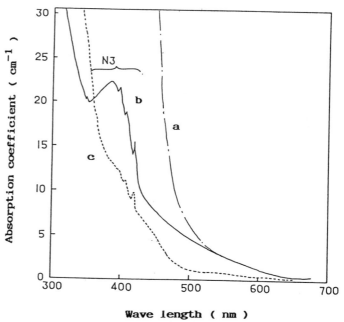

Figure 7. UV-Visible range absorption spectra for a, as grown crystal; b, (100) sector heat treated under high pressure at 1750°C for 50 hours after irradiation; c, (111) sector treated under the same conditions as b.

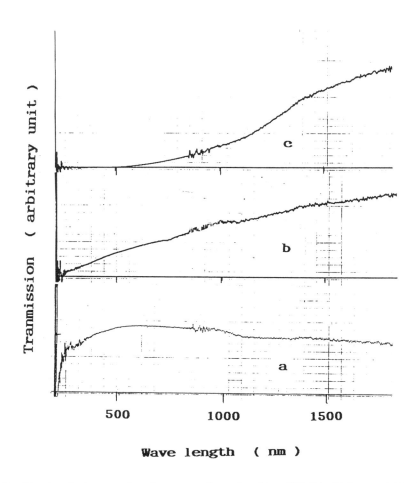

Figure 8. Typical transmission spectra in the UV-Visible range of a, white 12 micron thick; b, white 25 micron thick; c, black 37 micron thick polycrystalline diamond films.

Figure 9. SEM images of the film surface of a, white 12 micron thick; b, black 37 micron thick diamond films.

THE END OF THE BEGINNING

Russell Seitz
Center for International Affairs
Harvard University

I

"An Unctuous Substance Much Coagulated":
Opticks, Sir Is. Newton, 1710

When, in the course of advancing the state of the art, one slams into a material barrier to the construction of one's appointed gadget, it is customary, and at times mandatory, to drop to one's knees and pray to DARPA for deliverance. Deliverance in the form of the right stuff. Something superbly strong, something utterly transparent, something remarkably light, and with dielectric properties rivaling a perfect vacuum. Ideally, this something should also be bulletproof, better at conducting heat than a silver spoon, insoluble in boiling acid, radiation hard, non-toxic, and cheap. Well, historically, nine out of ten isn't bad for a start. Indeed, it's better than nothing.

So let me begin with some history. What we today call solid state physics began not as science but as technology. Victorian low technology to be exact. The first practical solid state electronic devices, demonstrated by Ferdinand Braun* at Leipzig on November 14, 1876, were based neither on theory nor on synthesis, nor on crystal growth. For in those days these things existed not. They were instead dug up, mined as lead ore. The performance of the galena (PbS) cat's whisker diode was marginal; it was rapidly superseded by the first, worse vacuum tube . So also, early infrared optics of rock salt gave way to synthetic crystals. But those early artifacts' performance demanded a physical explanation, and after a brief hiatus, in order for Willard Gibbs to break ground by inventing thermodynamics, the modern theory of solids arose to provide it. It all stemmed from the enterprise of explaining first the optical properties and then the electronic behavior of crystals found in rocks.

Today diamond, along perhaps with Iceland spar, remains the last optical material to be technically exploited as it is found in nature. It is a barbarous relic, a throwback to high technology's dim Neolithic past. For nowadays we are used to thinking about synthetic optical materials, like zinc sulfide or selenide, as being mature, as being just so much up market optical glass. They're stock items--you pay the money and the stuff shows up in big transparent slabs. That wasn't true even in 1970.

Then, as with diamond, there was essentially none to be had. The first measurements of the non-linear refractive index of ZnS had to be made on a sample of zinc ore from a Harvard museum--a pale green crystal of sphalerite whose optical quality then represented the state, not of the art, which was then non-existent, but of nature at her best.

So what can we do today when we're interested in a material that we are still learning to make as well as nature does? There are two maxims of the earth sciences that could be of service in the near term to the area of materials science that we are gathered here to discuss. One is: "The best geologist is the one who has seen the most rocks"; the other is more germane still: "Rocks are just ceramics that happen to have been made by God." Diamond is no exception, for today more than ever before, ceramics aren't just common clay.

So, naturally, there presently exist more shapes, forms, growth morphologies, surface textures, and degrees of optical quality in the diamonds of a good mineralogical collection than have, as of yet, been synthesized. Good, bad, and ugly, these varieties of natural diamond fall into two categories: the ones we have seen already synthesized and the ones that we will see. And the more we can find out about the natural history of the latter, the sooner we will see them synthetically reproduced.

---

* Who is today better known for his firm's electric coffee mill than for being the first man to violate Ohm's Law.

DIAMOND '87 = SILICON '38

- GROSSLY IMPURE

- HIGHLY STRESSED

- FULL OF ROCKS

- HIGH DISLOCATION DENSITY

- FULL OF COLOR CENTERS

- OVERPRICED (BY A CARTEL)

- ISOTOPICALLY IMPURE

- UNFIT FOR SEMICONDUCTOR USE

- RADIOACTIVE

SYNTHETIC DIAMOND '89–'98

NONE OF THE ABOVE

In this respect we enjoy a singular advantage. Unlike diamond, silicon just isn't found in the ground. But diamond is--presenting us, in advance, with a plethora of information about the pathology of its crystal growth that can teach us a great deal about what not to do. Here's a data base for the taking, and the Lord helps them who help themselves. So for God's sake, make the acquaintance of some mineralogists--a billion-year head start is not to be discounted lightly.

Given the abundance of carbon in the Earth's crust, one is staggered by the rarity of diamond. One wonders what would have happened to electronics if, by some quirk of nucleosynthesis, our sun were not a main sequence star, and that the world's supply of minable silicon ore were to be reckoned in ounces. Such a limited supply could never have sustained the sheer quantity of research that led to the development of silicon and the ensuing revolution in electronics and computing. Likewise, it is cautionary to reflect on what did happen, fifteen years ago, when the entire world's supply of optical-quality synthetic diamond could be easily accommodated in a volume of one cubic inch and the largest crystal yet grown could easily have been carried off by a fair-sized ant.

That half-centimeter crystal was of respectable quality, by natural diamond standards, that is, but compared to electronic grade silicon it was pathetic. Perhaps five orders of magnitude less chemically pure and three orders of magnitude more dislocation ridden. So it was with some trepidation that we exposed that tiny window, crudely mounted, to the full fury of the most powerful CW laser in the unclassified world, to a ten kilowatt CO_2 laser beam. One focussed to a scant millimeter in diameter and customarily used to slice thick steel plate like butter. With an average CW power density of a megawatt per cm^2 and a central power density an order of magnitude higher streaming through it for many seconds that little crystal barely got warm. My colleagues got a bit of a shock some weeks later, using a natural diamond of even more modest dimensions. They literally gave it their best shot.

PROTOTYPE WINDOWS

GORDON McKAY LABS

PASSIVE HEAT SINKS

10^8 WATTS/cm²

1973

AVCO-EVERETT

WATER COOLED

2×10^8 WATTS/cm²

1975

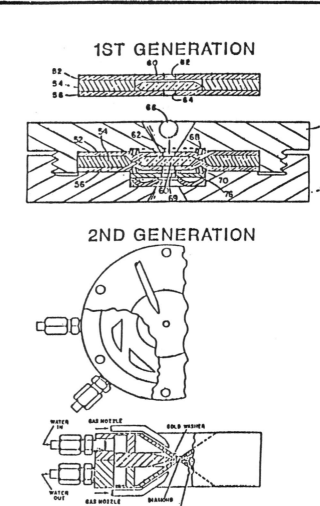

What my co-author Dairmid Duglas-Hamilton related in good dry Scottish prose in the pages of the <u>Journal of the Optical Society of America</u> (1-74) bears witness to the reality of the awesome thermal stress resistance parameter of the material in question: "At 5 KW a glow was observed where the laser beam entered the mount; at 10 KW a violent glow with ejected pieces of material occurred." Whereupon, I think prudently, Dairmid turned the laser off. But a surprise awaited him: "Inspection of the diamond showed it to be unharmed . . . After subsequent cleaning in <u>aqua regia</u>, no trace of damage was visible." No problem--the fireworks were nothing more, or less, than the 23% Fresnel reflection from the (as-sawn and unpolished) first surface of the window vaporizing the parts of the steel mount that got in the way. We were impressed.

But something of a comedy of manners then ensued. GE got its nice diamond back intact and AVCO, with DARPA's blessing, put a first-rate heat transfer team to work building a water-cooled mount that worked nicely. Operating at a Reynolds number of 50,000, it should have. But that team was new to diamond, so they delegated the task of producing some fifty small natural diamond windows to a New York lapidary firm. One that was then unfamiliar with optical physics. (This was before the days of the Pioneer Venus infrared radiometer window.)

THERMAL STRESS RESISTANCE
FIGURE OF MERIT

	S_f (MPa)	E (GPa)	$a \times 10^{-6}$ /°K	K(W/M°K)	$R = \frac{S_f(1-N)K}{a E}$ (W/M)
GLASS	90	52	11.4	0.62	80
SILICA	200	71	0.55	0.8	350
SILVER	100	70	20	406	1.8×10^4
SAPPHIRE	440	405	6.7	50	6×10^3, 300°K 1.8×10^4, 30°K
DIAMOND	200	1000	0.8	2,000	3.6×10^6, 300°K 3.6×10^8, 77°K

SEVEN ORDERS OF MAGNITUDE BETTER THAN GLASS

And what was sent for testing in 1975 was deplorable, a random mine-run assortment, mostly of poor IR transparency. Only two of that very expensive lot were anywhere near the quality of what we tested in 1973.

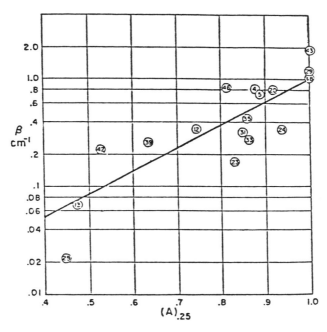

Distribution of absorbtion coefficient at 10.6 μm (ß) versus absorbtion coefficient at 0.25 μm (A) for type I and type II diamonds.

Evidently DARPA was not amused. The funding stopped and that early chapter in diamond power optics drew to an abrupt close. Few large optical elements could be shrunk down to the size of affordable natural diamond and scaling synthetic diamond up in size beyond a quarter of an inch was, or so we thought, beyond the limits of state of the art. We were wrong.

For, meanwhile back in Johannesburg, some clever crystal growers at DeBeers were doing what all astute employees of the diamond cartel do when they synthesize centimeter-size synthetic gem diamonds--not telling anyone. So progress seemed retrograde. And meanwhile, back in the Soviet Union, the proprietors of the world's largest high-pressure press were either working on the Five-Year Plan to unsuccessfully make metallic hydrogen or telling each other Polywater jokes at Derjaguin's expense. It was not a brilliant decade on the whole.

SIZE HISTORY

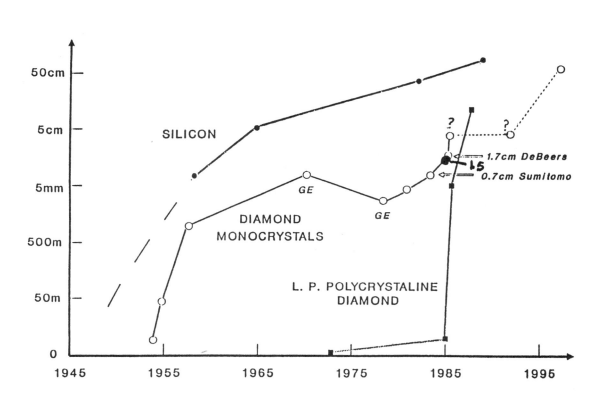

There is a clear moral to this episode: this area of research requires constant vigilance. Read the data on natural diamond with caution. For what we call intrinsic, DeBeers calls "D-flawless." And the diamonds they have provided to researchers over the years have rarely been of that quality. So the data base on the physical properties of diamond has (naturally) been heavily skewed away from intrinsic values. The Earth's mantle is not a clean room and explosive vulcanism is not very respectful of lattice quality. So only a very few natural diamonds approach intrinsic perfection. Yet these benchmark crystals do exist.

In Britain there is one about a quarter of an inch long whose Lang topograph shows just four dislocations! We could use more like it--I think it should be cloned. Likewise, that unlikely center of solid-state research, Harry Winston & Co., admits rather reluctantly that they handle some West African diamonds with undetectable nitrogen contents. But you won't find them in the literature--they cost too much to lend to scientists. The biggest of the best of these (130 carats) was sold last year for a megabuck a gram in the rough.

STATE OF THE ART

	ISOTOPIC PURITY	DISLOCATION DENSITY	SIZE	OPTICAL ABSORPTION	CHEMICAL PURITY	RADIATION DAMAGE
SILICON	94%	10^2	~60cm	$<10^{-4}$	.999999999	NIL
NATURAL SAPPHIRE	99.97	$10^4 - 10^{10}$	~5cm	$>10^{-2}$	.995	$10^7 - 10^8$
SYNTHETIC Al_2O_3	99.97	10^3	~100cm	$<10^{-4}$	.99999	NIL
NATURAL DIAMOND	98.9	$10^1 - 10^{12}$	$<5cm$	$10^2 - 10^{-2}$	.99 - .9999	~ 10^8
INTRINSIC MONONUCLIDIC ^{12}C	>99.9999	~10^2	$>10cm$	$(?)10^{-4} - ^{-7}$	.999999	NIL

DIAMONDS ARE FOREVER; BUT,
FOREVER x 100 MILLIRAD/YR = LOTS OF RADS!

And don't overlook yet another artifact in the data base--all of the measurements are, of necessity, on small samples. Looking at thermal conductivity graphs we see higher cryogenic peaks for sapphire and nearly monoisotopic lithium fluoride than for diamond--what is going on here? That's not in keeping with theory. On reading the fine print one discovers that the highest recorded thermal conductivity for diamond was for the largest specimen ever measured--just 12 millimeters, versus multi-centimeter bars of Al_2O_3 and Li^7F, whose size translates into higher values of K, despite lower values of theta Debye. The deHaas- Casimir scaling of thermal conductivity with free path is evidently very real. What are its practical limits? I suggest that we may see multicentimeter crystals or thick films grown long before we are able to persuade the Warden of the Tower of London to dunk the five-centimeter type IIA that adorns the scepter of the realm into liquid nitrogen. He is not obliged to satisfy our curiosity. But the Royal Society is working on the case.

But as the eighties rolled around diamond-like carbon films became more and more diamond-like. Two and a half centuries after Newton's wise conjecture that diamond was "an unctuous Substance much coagulated"**, things finally gelled for real, and a couple of paradigms started to shift. The FEL started to evolve, laser diodes underwent a population explosion, VLSI device densities grew at a scorching rate. And across the nation, applied physicists in many disciplines started to demand material solutions, solutions to problems reckoned in too many kilowatts per too few square centimeters.

** Sir Isaac's revolutionary view was a marked improvement over that of classical authorities like Aristotle and Theophrastus. They stoutly maintained that diamond was, Q.E.D., just another polymorph of water, like ice, quartz, or pearls. This is known as the wisdom of the ancients.

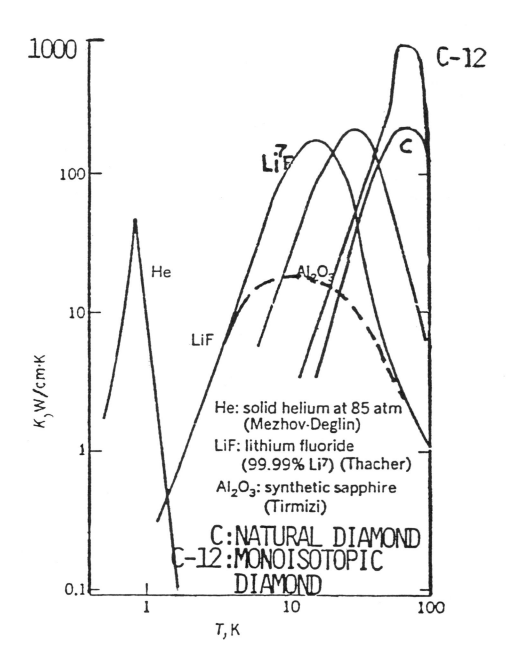

He: solid helium at 85 atm
(Mezhov-Deglin)
LiF: lithium fluoride
(99.99% Li⁷) (Thacher)
Al₂O₃: synthetic sapphire
(Tirmizi)
C: NATURAL DIAMOND
C-12: MONOISOTOPIC
DIAMOND

And in laboratories across the world, a lot of people started to believe their eyes--equilibrium thermodynamics notwithstanding, low-pressure diamond synthesis had gone from clearly irreproducible to almost unavoidable. There is more: high-pressure single crystals have reached 1.7 cm in size--ten carats, and getting bigger by the month!

The hard problems of chemical and isotopic purification are at long last being attacked on all fronts. And high time, too. Just as Phosphorous is a fatal toxin even at ppm levels where electronic silicon is concerned, so likewise we have to banish nitrogen if viably transparent, and eventually intrinsically transparent, diamond is to be achieved. Even when chemical purification has been perfected, that ubiquitous relic of the solar system's formation, carbon 13, will remain to bedevil the dynamics of lattice vibration and ballistic phonon propagation, especially affecting the optical branch. When I asked last year how many nines pure was the metal in which most abrasive diamonds are grown, the answer was "Maybe two." My inquiry into the nitrogen content of the best methane on the market met with an answer that explained much:

"I don't know." Last, and worst to report, the effort to exorcise that strong optical poison from the cylinders in which we all take delivery of the finest kind of CH_4, consists in first filling them with air and finally, connecting them to a roughing pump. Considering the care lavished on silane (SiH_4), the mind is repelled.

ECONOMICS

	HIGH PRESSURE	LOW PRESSURE
YIELD PER PRESS/REACTOR	5KG/YR	175 KG/YR
CAPITAL COST	~$5,000,000	~$500,000
INTEREST & AMORTIZATION/KG	~$100,000	~$285

	COSTS	CONSUMPTION
GOLD	$200/cm³	50 TONS
DIAMOND	$35 - 100,000/cm³	8 TONS

What are we to make of all this? Diamond research clearly differs from both earlier cycles of obsolescence and substitution in materials innovation. And also from the other instances of sudden fashion in contemporary materials science. The former have been heuristic, but the process of climbing the ladder of the periodic chart from germanium to silicon ends

DIAMOND APPLICATIONS

WINDOWS & OPTICS

FLIR	RADIOMETERS
GYROTRON	• PIONEER VENUS LANDER
FEL	TOKOMAK
HEDI	WELDING ROBOT
OPTICAL ARMOR	SURGICAL LASER
• PLANES TANKS & GRUNTS	ASW ILLUMINATION

DIAMOND APPLICATIONS

	# UNITS
RAZOR BLADES	10^{10}
IC SUBSTRATE	10^8
DIODE ARRAY	10^7
ARMOR VISOR	10^6
WELDING OPTICS	10^5
HEDI WINDOW	10^3
PLASMA TORUS ACCELERATOR	10^2
NEUTRINO DETECTOR	10^1
GIGAWATT FEL	5×10^0

abruptly at diamond--one step further up the tetrahedral rungs and you fall off. The latter, the niche of fashion presently occupied by High T_c superconductivity, must await changes in both infrastructure and electronic design philosophy. Being both linear in its thrust and limited in its sphere of application, I think High T_c will complement rather than compete with the optoelectronic applications of diamond. It may seem to be the Ace of Spades, but we seem to be holding the rest of the deck. Besides, what bloody use is a superconducting razor blade?

II

Ne Plus Ultra

In considering the scope of its impact on technology both high and low, remember that, just as banishing electrical resistivity opens new vistas in electrical circuit design, so does banishing the bulk of thermal resistivity transform the art of heat transfer. If electrical engineers had only iron wire to work with, where would we be? Today thermal control engineering is stuck with copper as the epitome of heat conduction--and look where we are. Clearly, we aren't going to stick around in this regime. For diamond affords us the solid equivalent of a liquid heat pipe--and it is but <u>one</u> of our face cards. And the Jack of Diamonds throughout history has <u>always</u> been a trump.

I therefore submit that the product of all the many physical superlatives embodied in the properties of diamond, not just electronic, but optical, thermal, and mechanical, will effect and enable a broad spectrum of advances in technologies, both high and low. One that will be clearly broader in technical scope and probably of a macroeconomic depth exceeding those due to the high T_c revolution. In reflecting on what has been termed the Woodstock of Physics, one must remember that, as in the original event, not only was the crowd both huge and ecstatic, but that the band was rather small, and few of the tunes they played were long remembered. Do not misunderstand me--high T_c research is a noble calling, and a hefty wedge of the economic pie of electronics will reward its fruition. But our confection, equally half-baked, is far larger in its ultimate diameter. Besides, we share a common motto with our esteemed colleagues: "If you can't stand the cold, get out of the kitchen."

HISTORY OF MATERIALS

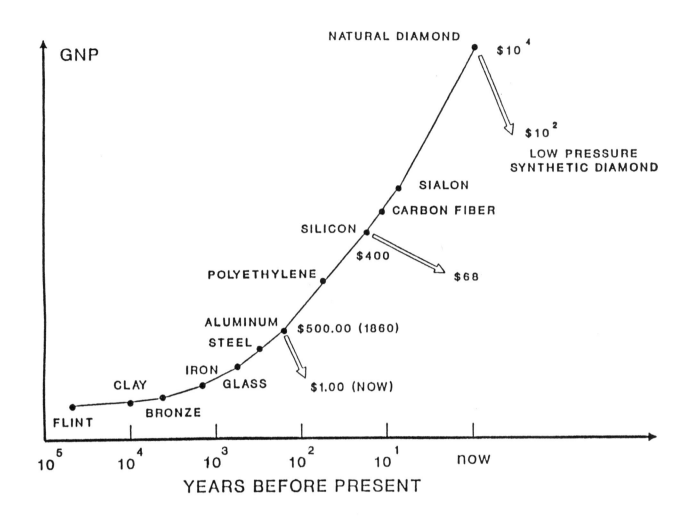

I want to briefly discuss the prospect of how cheap diamond can ultimately get, and how soon values close to its intrinsic properties may reasonably be achieved. Basically, it will be relatively easy to make tons of gray, opaque, ugly, but mechanically excellent diamond for less than 500 bucks per kilo. But it will be relatively hard to make kilos of diamond as good as intrinsic single crystal silicon in five-centimeter sizes for less than a hundred times that price.

By relatively hard, I mean that the integrated cost of the R&D cycle over a five-year term may approach 1% of the money already spent in bringing silicon up to speed, or 10% of the total expenditure on GaAs over the last decade. Cracking this case, raising diamond up to the level of silicon's perfection, is going to cost a fortune.

So how on earth can I justify spending a small multiple of 10^8 dollars? The answer is simple. Solve such problems, just as they were solved in silicon's case, and a plenum of unanswered prayers would abruptly be fulfilled. And, incidentally, large sectors of the market for silicon, GaAs, and even InP, would fall by the wayside, victims of the fulfillment of DARPA's unspeakable ultimate mandate. Its real task is to make things obsolete.

Which is where diamond comes in: there is something brutally unique about carbon and the other elements of the first row of the periodic chart. They inhabit the <u>ultima thule</u> of solid state physics. Because, I repeat, there is <u>no</u> row zero or minus one.

So as surely as in the slow ascent from cat's whisker diodes made of lead ore, to germanium to silicon and beyond, each new material has both opened new vistas for the engineers and has gobbled up the market share of its older subordinates; so likewise the more robust adamantine solids, beta silicon carbide, boron phosphide and the like, will overrun the lesser III-V compounds and be displaced in turn by diamond and boron nitride.

For we all live in the material world. And one of its iron laws is starkly Darwinian: pick a figure of merit--any figure of merit--and apply it to any electrooptical material. Then plug in the numbers for diamond--either bad natural or good synthetic. And behold: natural selection at work. In the fight for funding on the basis of intrinsic merit that defines the food chain of materials science, only one of God's creations can emerge triumphant, red in tooth and claw. It is the macromolecule at the top of the chart that we call diamond. It's where an evolutionary trend in materials science culminates. And where a newer and more elegant epoch in the evolution of applied physics and high technology starts. Consider some problems that already exist.

One example of a set of geometrically dictated engineering problems that require monolithic diamond as a solution is tip loading. Whenever a high-power density coincides with a component that has to be tapered to an edge or a point that power density rises toward infinity and things tend to get literally blown away. Axicon mirrors in lasers, hypersonic nose cones and lead edges, and the working end of pulsed power systems like the compact Torus plasma accelerator all present the same syndrome. The first or last ten centimeters have to face up to gigawatts of power or kiloteslas of brute magnetic force. Generally not much can be done about the last few centimeters. It is doomed to be vaporized. But oftentimes a design compromise just this side of coming to a point will prove enabling. The unique combination of thermal diffusivity, transparency and great mechanical strength that diamond affords opens up an unexplored area of solid materials co-existing with powerful energy fluxes and shock waves.

EQUIPMENT HISTORY

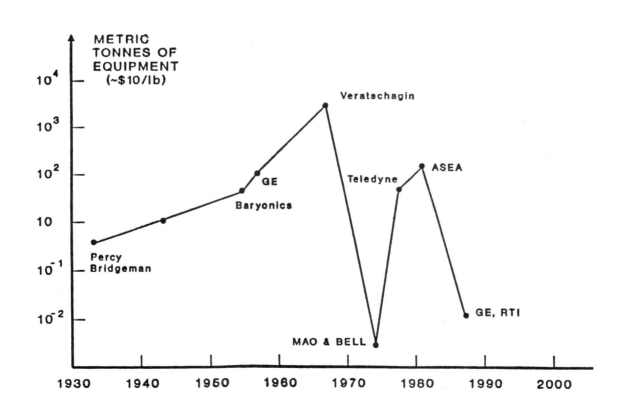

The identity of pressure and temperature has been with us since Percy Bridgeman used 7 kilobars instead of heat to hard-boil an egg at room temperature. So it is important to reflect on the profound difference between the response to dynamic loading of customary engineering materials and adamantine solids. Subject the former to pressures of a hundred thousand atmospheres and they flow like water. But to subject diamond to 10 gigapascal is to do it a favor. Surprise: you've just put it back inside its thermodynamic pressure stability field. Stress it further and, eventually, it will shrug stoically into first lonsdalite and, ultimately, metallic carbon. But it's the first megabar of grace that counts. It puts diamond in the curious position of being stronger than conventional explosives conventionally detonated. Curious in the sense that their detonation velocities are slower than the speed of sound in its crystal lattice. So, for the first time, we have microseconds of leisure in which to do something about the energy that a point-blank detonation wave deposits at the surface, a surface that will still be there instead of having become a part of the plasma that hit it.

Now, low-temperature ballistic phonon transport does some mighty Peculiar things to the reference frame Hugoniot curves of both diamond hammers and anvils--the latter can achieve thermal equilibrium through centimeter thickness in less than a microsecond; fire a small diamond crystal from a light gas gun at a large, cold block of diamond at the grossly subsonic velocity of 7 km/sec and instead of the customary puff of plasma you might get two lukewarm pieces of carbon and a lot of noise. Put something in between and it will experience both a huge pressure excursion--and a phenomenally fast quench rate. I note that this alien thermodynamic regime has never been visited before and we really don't know what metastable phases live there. It will be quite different from the nuclear explosive derived 10^9 bar environment and, obviously, far easier in terms of both access--and bringing the sample back alive.

III

The Hard Way

When it became evident to microwave technologists that GaAs was worth developing, both corporate and federal laboratories, here and abroad, began to commit long-term funding to the enterprise. What was known from the start was that then current silicon growing techniques would not work. Yet in the early sixties the first GaAs varactors somehow appeared--they were based on home-grown small crystals of a quality that would be virtually unsalable in today's market. But they were the state of the art. Warts and all, they broke out of the performance envelope of silicon and irrevocably changed the world's view of what microwave technology could accomplish. Today, likewise, we are already seeing remarkable devices, both optical and electronic, being produced from a few small bits of crude material, both natural and synthetic. And we are witnessing rates of improvement in size and quality that are proportional to the expansion of research effort and funding. Over the past few years, that level of investment has increased relative to a decade ago by more than an order of magnitude--to about 10^7 dollars per year.

Insofar as this is about two orders of magnitude less R&D funding than is still lavished on silicon annually, one wonders at the progress that is occurring. And one wonders also what will happen as funding matures and the communities' expertise expands. One paradigm concerns the silicon industry, the economic history of its technology has been quite thoroughly researched. Erik von Hippel of MIT's Sloan School has charted the course of the enabling developments that led to VLSIC-grade silicon and beyond. A course punctuated by a half dozen episodes of innovation. The last of these was the rediscovery of silicon's non-electronic thermomechanical properties and hence its utility in high-power optics.

This should cause us to reflect. For this meeting signifies not only the end of diamond's infancy as an optical material, but the beginning of the transition of free-standing silicon I.C. technology from maturity to thermal senescence. Good old Silicon, it was nice while it lasted. But the inexorable progress of optoelectronic device technology is catching up with its intrinsic physical limitations. Having at last mastered the full spectrum of what can be done with silicon, the designers have begun to do things to it. And it is accordingly feeling a world of hurt. It's high time to get ourselves a new design envelope--one with broader horizons and a higher ceiling--about three orders of magnitude worth. One that can sustain into the next century a further inflation of performance--and of hope.

That hope was once solely the province of solid state theoreticians to whom a material's intrinsic properties are more real than what we can measure and observe. Their mathematical constructs exist outside the time in which technology is embedded. So they had the certain and authentic knowledge of what silicon would eventually be as long ago as the forties.

But none would then have dared to extrapolate from the best silicon money could then buy (which was just 99% pure, with some crystals visible to the naked eye) to the silicon crystals of today, crystals the size of railroad ties ten nines pure with fewer dislocations than a grain of sand.

It's not what they expected, but it's what they got, starting from scratch. It took some hundreds of thousands of man years and tens of billions of dollars. It was, in retrospect, the best research investment in the history of technical civilization. It has paid better than a hundred to one--literally trillions of dollars have accrued to the world's GNP in consequence of the silicon revolution.

But there's more--we now have, in direct consequence of the electronic revolution and the need for new materials that it spawned, something that did not exist a generation ago: materials science. An integrated interdisciplinary enterprise that can draw on a plenum of human, technical, and informational resources. An enterprise of discovery and development within which diamond's evolution will be far better nurtured than silicon's ever was. We can accordingly anticipate that the acceleration of history that science brings will continue and that the next chapter, chronicling the emergence of the last and best element of the periodic table's fourth group, will be written at a faster pace than those preceding.

Just how fast will depend in large measure on our capacity to remember what has been learned already about not just diamond, but its isoelectronic shadow, boron nitride, and its half-sibling carbon silicide. I reverse the customary usage with more than rhetorical intent--our understanding of that material is far deeper than our practical knowledge of carbon carbide, but shallow indeed compared to our mature understanding of how to manipulate the band gap of silicon silicide. And Lord knows that that knowledge has been worth the knowing.

INITIAL DOD/SDI NEEDS

SIZE REQUIREMENT: 5cm

- 3 X BIGGER THAN DeBeers
- 3 X BIGGER THAN SUMITOMO
- 10^1-10^2 THICKER THAN RTI

THEORETICAL WORK TO BE DONE

- SURFACE INTERACTION OF PHONONS
- ^{12}C-^{13}C BAND GAP DIFFERENTIAL
- C/SiC INTERFACIAL PHENOMENA
- $^{28}Si^{12}C$ & OTHER MONONUCLIDIC SOLIDS

- INTRINSIC ABSORPTION
- SRS - SBS
- BALLISTIC PHONON TRANSPORT
- FREE PATH LIMIT/SECOND SOUND

But it won't be a simple matter. Organic chemistry is nature's way of telling us that beneath the very simplicity of diamond physics is a complex phenomenology. One that gives rise to more geometric and thermodynamic variations and imperfections in the crystal lattice and surface than silicon researchers ever faced. The dream and the nightmare are one--to make intrinsic silicon, you just take away the chemical and physical impediments to its automatic assembly via a first order phase transition at a freezing interface. To get carbon to assemble into diamond in the same straightforward and thermodynamically foolproof fashion will require an assault on the gates of Hell itself--four thousand degrees Kelvin and a respectable fraction of a megabar.

It's a long way from that regime to the tenuous plasmas of non- equilibrium growth. And in between there is a lot of territory to be explored. I think the explorers should succeed; their equipment is superb--the armchair has come of age. Today we can sit down at our computers and model and simulate both equilibrium and non-equilibrium interfaces in crystal growth in ways that were undreamed of in silicon's infancy. For it is only with silicon's technical maturity that mankind has been endowed with the gift of computational power.

Each passing year puts newer, more elegant, and less pronounceable spectroscopies and diagnostic tools at our disposal. Some of them even work. And as this evolution continues, before long we may, to the vast satisfaction of the solid state theoreticians and the great discomfort of the world's diamond miners, begin to say that we know what we are doing. I fervently hope that NMAB, DSB, and OTA will rise to the occasion together with the Primes. For if they fail to be of good courage (and deep pockets), we are all going to end up like the star of a very short film called Bambi Meets Godzilla, abruptly and unceremoniously stomped into obsolescence.

For when the cutting edges of U.S. and Japanese high technology clash on this front, both had better be of adamant. Theirs already is, and if ours isn't, we won't even get a chance to parry. For this much is already certain. Diamond's applications will not be limited to America's defense, and many new markets, around the world, will be the lawful prey of the first nation to industrially deploy this extraordinary material.

Diamond films for improving survivability of thin film metal mirrors against x-ray

James Tsacoyeanes and Tom Feng

Perkin-Elmer Corporation
100 Wooster Heights Rd., Danbury, Connecticut 06810

ABSTRACT

Our theoretical results have demonstrated the usefulness of diamond films for improving the survivability of thin film metal mirrors against intense x-ray. Essentially, the low-Z, high thermal conductivity diamond film acts as an efficient heat sink, thus reducing the temperature in the metal film. Specifically, for an aluminum/fused silica mirror, our analysis has shown that interposing a 1-μm thick diamond film between the thin film aluminum and SiO_2 substrate increases the aluminum melt fluence--from a 1 KeV black body, 10 ns x-ray pulse--by a factor of 6. In addition, we have found that for pulse widths on the order of about 10 ns, a diamond film thickness of about 1 μm appears to be most effective in reducing the aluminum temperature.

1. INTRODUCTION

Thin-film coated metal mirrors have a variety of surveillance and communication applications. However, the survivability of such thin-film optical mirrors with respect to nuclear radiation is of major concern. This is especially true of aluminum coated, fused silica mirrors because of the low melting point of aluminum (660°C). In the early 70's a concept was developed for hardening the optical mirror coatings against intense x-rays.[1] The idea was to interpose a low-Z, high thermal conductivity film between the relatively high-Z metal film and substrate. In effect, the low-Z film will act as an efficient heat sink, thereby redistributing the deposited heat load and lowering the temperature of the metal film.

In general, beryllium is used for this purpose because of its very low x-ray absorption and relatively high thermal conductivity. However, due to its extreme toxicity, special precautions must be taken in depositing the beryllium film. It appears to us that diamond--which has a low-Z value (6) and the highest thermal conductivity--is an ideal material for such an application. The reason why diamond has not been investigated, or even considered for such an application, is mainly due to the fact that until recently diamond could not be obtained in a thin film form. However, recent breakthroughs in the low-pressure synthesis of diamond by chemical vapor deposition (CVD) techniques have made it possible to deposit thin films of polycrystalline diamond on a variety of substrates.[2-7] Although the films are polycrystalline in microstructure, the thermal conductivity of the present state-of-the-art diamond films has been shown to have values approaching that of type I natural diamond.

In this paper computer simulation results of using diamond for this type of application will be presented. Specifically, it will be shown that an aluminum coated, fused silica mirror with a layer of diamond in between significantly improves the survivability of the mirror against intense x-rays.

2. BACKGROUND

In general, higher than band-gap photons interact with the solid by one of three processes. They are: 1) the photoelectric effect, 2) the Compton effect, and 3) electron-positron pair formation. Since the threshold of electron-positron pair production is about 1 MeV and the cross section for Compton scattering is comparatively low, in the energy range of interest (less than about 10 KeV) the primary absorption mechanism is due to the photoelectric effect. The cross section for the photoelectric effect is a strong function of the wavelength, λ , and the atomic number, Z, of the material. Away from the absorption edges, it varies approximately as the third power of λ and the fifth power of Z.

As the energetic electrons generated by the x-ray photons travel through the solid, it iteracts with the lattice electrons. As a consequence, the lattice electrons are excited to produce phonons, so that the material heats up. Unless this heat is promptly removed, the temperature in the material will rise until it melts or other failure mechansim occurs. The x-ray intensity is attenuated by the material, and it decreases exponentially as a function of the depth into the material. It obeys the standard Beer-Lambert Law:

$$I = I_0 \exp(-\mu\rho x),$$ (1)

where μ is the mass absorption coefficient and ρ is the density of the material.

The effectiveness of the film as a heat sink depends primarily on its x-ray absorption and heat conduction characteristics. The amount of x-ray absorption by the material can be judged from the deposition depth, X, defined as

$$X = 1/\mu\rho .$$ (2)

X is the thickness needed to attenuate the x-ray intensity to 1/e of its incident value. The heat conduction, or dissipation, capability of the material can be judged by the thermal diffusion length, L, which is defined as follows:

$$L = (\tau D)^{\frac{1}{2}},$$ (3)

where τ is the x-ray pulse length and D is the thermal diffusivity given by

$$D = k/\rho C_p,$$ (4)

in which k is the thermal conductivity and C_p is the specific heat at constant pressure.

The spectral distribution of the x-ray pulse can be approximated by a black body at a specific temperature and can be determined from Planck's law.

$$E(\lambda,T) = \frac{C_1}{\lambda^5 \, [\exp(C_2/\lambda T)-1]},$$ (5)

where $E(\lambda,T)$ is the emitted power, λ is the wavelength, T is the temperature, and C_1 and C_2 are constants.

3. ANALYTICAL APPROACH

The mirror configuration chosen for this study is representative of a typical aluminum-coated, reflective metal mirror. It consists of a fused silica substrate coated with a 75 nm thick aluminum film. To this basic design, a diamond layer is interposed between the aluminum film and fused silica substrate. Figure 1 shows the two mirror configurations investigated here.

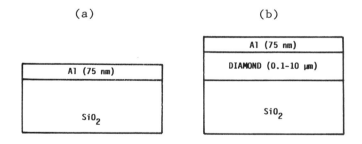

Figure 1:   Two thin-film metal mirror configurations: (a) without a diamond interfacial layer and (b) with a diamond interfacial layer.

In our analysis the incident x-ray source was chosen as a 10 ns pulse with a black-body spectrum corresponding to a temperature of 1 KeV. The total power, P, deposited in a layer of thickness x is found by integrating over energies and distance as follows:

$$P = \int_o^x \int \frac{C_1\mu\rho \, \exp(-\mu\rho x')}{\lambda^5 \, [\exp(C_2/\lambda T)-1]} \; dx' d\lambda.$$ (6)

The solution of Eq. (6) becomes the source term for solving the one dimensional thermal diffusion equation.

$$\frac{\partial^2 T}{\partial x^2} - \frac{1}{D}\frac{\partial T}{\partial t} = \frac{P}{k}. \tag{7}$$

To determine the temperature distribution in the various layers of the mirror as a function of time and depth during and after irradiation by the x-ray pulse, we solved Eq. (7) numerically using a combined absorption and finite-difference thermal (SINDA) program. It should be noted that in Eq. (7) both the diffusivity, D, and thermal conductivity, k, are functions of temperature. The experimental values of the thermal conductivity and the specific heat at constant pressure needed to calculate the diffusivity of the three materials were obtained from tables compiled by Touloukian et.al.[8] For diamond, however, values of the thermal conductivity above 127°C were obtained by using the expression[9]

$$k = 3 \times 10^5 \ T^{-1} \ (Wm^{-1}K^{-1}). \tag{8}$$

Values of the thermal properties of diamond used in this analysis were those for single crystal, type IIa, natural diamond. Although this makes the results somewhat hypothetical, they allow an assessment of the potential of diamond films for radiation damage applications. In addition, present state-of-the-art diamond CVD technology is already producing polycrystalline films with thermal conductivity approaching that of natual diamond. For example, Sawabe and Inuzuka[5] and Ono et.al.[10] have shown that CVD deposited polycrystalline diamond films have thermal conductivity as high as 10 Wcm^{-1} K^{-1}. Such a value is comparable to that for type I natural diamond, although less than the best type IIa natural diamond (20 Wcm^{-1} K^{-1}).

4. RESULTS AND DISCUSSION

The effectiveness of interposing a 1 μm thick diamond film in hardening a thin film (75 nm) aluminum mirror against intense x-rays can be clearly seen from Fig. 2. Shown in Fig. 2 are the temperature profiles as a function of time of the various layers of the mirror during and after the 10 ns x-ray pulse. The incident x-ray fluence was chosen to be 0.13 cal/cm^2, corresponding to the melt fluence for aluminum without a diamond film--i.e., the fluence needed to raise the temperature of the aluminum film to its melting point (660°C). In this calculation, values of the temperature of the aluminum and diamond films were taken at the midpoint between the two interfaces, whereas for the SiO$_2$ substrate it was taken at a point 1 μm below the diamond-SiO$_2$ interface.

During the duration of the x-ray pulse, the temperature in all layers rises rapidly due to the dissipation of the x-ray energy in the materials. After the pulse ends, the aluminum film will initially cool down and the diamond film heat up until a quasi-equilibrium is reached. Because of the good thermal coupling between aluminum and diamond, this thermal equilibration will occur in the first few nanoseconds. For example, at 200°C, the thermal diffusion lengths for a 10 ns pulse are 925 and 1636 nm for aluminum and diamond, respectively. Since the SiO$_2$ substrate is orders of magnitude thicker than the diamond layer and the fact that it has a very low thermal diffusion length (e.g. 91 nm at 200°C for a 10 ns pulse), the presence of the diamond film has almost no effect (within a few degrees) on the temperature profile of SiO$_2$. It is important to note that as time passes, there will be a thermal equilibration among all the layers. After the initial cooling, as indicated in Fig. 2, the aluminum film will gradually warm up as heat flows from the hotter fused silica into the cooler diamond and aluminum. However, due to the poor thermal conductivity of SiO$_2$, this process will occur at much longer time scale than that shown in Fig. 2. Eventually, a thermal equilibrium will be reached between the layers, and the mirror as a whole will asymtotically cool down to ambient temperature.

As can be seen, there is a significant difference in resulting temperatures for the two different mirror configurations. For a fluence of 0.13 cal/cm^2, the aluminum temperature reached at the end of the pulse for a standard mirror, corresponding to Fig. 1a, is 660°C. Adding a 1 μm thick diamond layer reduces this temperature to 170°C, well below the melting temperature of aluminum.

Next, we increased the incident x-ray fluence for the 1μm diamond mirror until the aluminum begins to melt. This fluence was determined to be 0.77 cal/cm^2, which is six times the melt fluence without the 1 μm diamond layer. The temperature profiles of the various layers under this fluence condition is shown in Fig. 3.

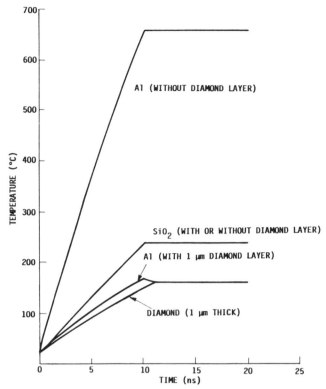

Figure 2. Effect of interposing a 1 μm thick diamond film between the aluminum film and the fused silica substrate. The x-ray source was a 10 ns pulse with a black body temperature of 1 KeV and a fluence of 0.13 cal/cm².

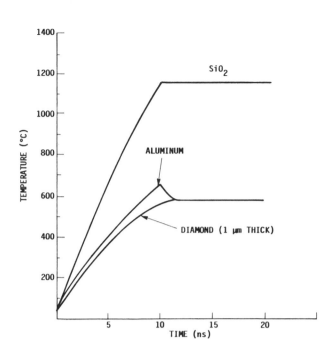

Figure 3. Temperature profiles of the various layers under an aluminum melt fluence condition of 0.77 cal/cm². On a much larger time scale, the SiO_2, Al, and diamond will gradually approach the same temperature.

Figure 4 shows the effect of the thickness of the diamond layer on the temperature in the aluminum film. In this analysis the incident x-ray fluence was assumed to be 0.13 cal/cm², and the aluminum temperature was taken at 10 ns, i.e. at the end of the pulse. It is evident that very little additional advantage can be gained by increasing the diamond thicknesses beyond about 1 μm. This is because for a 10 ns pulse, the thermal diffusion length of diamond is on the order of 1 - 2 μm, so that on a nanosecond time frame very little heat will flow beyond a depth greater than about 1-2 μm.

5. CONCLUSIONS

Our analysis has shown that interposing a thin diamond film between the reflecting metal film and the substrate of a metal mirror significantly improves the survivability of the mirror against intense x-rays. Essentially, the low-z, high thermal conductivity diamond film acts as an efficient heat sink, thus reducing the temperature in the metal film. Specifically, for an aluminum/fused silica mirror, incorporating a 1-μm diamond film will increase the aluminum melt fluence from a 1 KeV black body, 10 ns x-ray pulse by a factor of six. In addition, for pulse widths on the order of about 10 ns, a diamond film thickness of about 1 μm appears to be the most effective in reducing the aluminum temperature.

6. ACKNOWLEDGMENTS

We would like to thank Roger Paquin and Alan Ganz for their helpful advice and stimulating discussions during the course of this work.

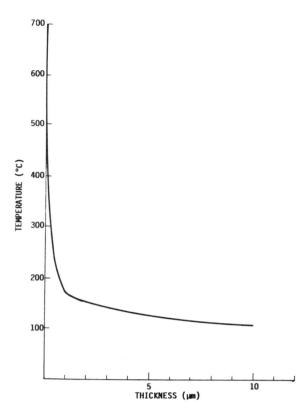

Figure 4.   Effect of thickness of the diamond layer on
the temperature of the aluminum film. A fluence
of 0.13 cal/cm², corresponding to the melt fluence
for aluminum without a diamond interfacial layer,
was used.

<u>7.   REFERENCES</u>

1.   G. R. Wirtenson, "Nuclear survivability and hardening of the optical coatings," Lawrence
Livermore National Laboratory Report No. UCRL-15738, (Dec. 1985).
2.   D. V. Fedoseev, V. P. Varnin and B. V. Deryagin, "Synthesis of diamond in its
thermodynamic metastability region," Russ. Chem. Rev. <u>24</u>, 435 (1984)
3.   M. Kamo, Y. Sato, S. Matsumoto, and N. Setaka, "Diamond synthesis from gas phase in
microwave plasma, J. Crystal Growth <u>62</u>, 642 (1983)
4.   K. Suzuki, A. Sawabe, H. Yasuda, and T. Inuzuka, "Growth of diamond thin films by
dc plasma chemical vapor deposition, "Appl. Phys. Lett. <u>50</u>, 728 (1987)
5.   A. Sawabe and T. Inuzuka, "Growth of diamond thin films be electron assisted chemical
deposition and their characterization," Thin Solid Films <u>137</u>, 89 (1986)
6.   R. Messier, A. R. Badzian, T. Badzian, K. E. Spear, P. Bachmann, and R. Roy, "From
diamond-like carbon to diamond coating," Thin Solid Films <u>153</u>, 1 (1987)
7.   A. Feldman, E. Farabaugh, and Y. N. Sun, "Diamond, a potentially new optical coating
material," <u>Proc. of Conf. on Laser Induced Damage in Optical Materials</u>, Boulder, CO (1987)
8.   Y. S. Touloukian, R. W. Powell, C. H. Ho, and P. G. Klemens, "Thermophysical Properties
of Matter, "Vol. 1 (1970)
9.   J. E. Field, "The Properties of Diamond," p. 13, Academic Press, New York (1979)
10.   A. Ono, T. Baba, H. Funamoto, and A. Nishikawa, "Thermal conductivity of diamond
films synthesized by microwave plasma CVD," Jap. J. Appl. Phys. <u>25</u>, L808 (1986)

SESSION 6

Diamond Optical Applications

Chair
K. V. Ravi
Crystallume

Applications of diamond in optics

M. Seal and W.J.P. van Enckevort

Drukker International B.V.,
20 Beversestraat, Cuyk 5431 SH, The Netherlands

ABSTRACT

This paper reviews existing and new applications of single crystal diamond, both natural and synthetic, in optical science. The traditional application is as transmissive components, making use of the very wide spectral transmission range, high thermal conductivity, and chemical inertness of diamond. Diamond windows for corrosive environments are well known; diamond surgical endoscope components are under development; and the use of sharpened diamonds as combined surgical cutting instruments and light pipes for internal illumination of the edge is commercial reality. The superb ability of diamond to conduct heat, combined with its very low thermal expansion coefficient makes it suitable for the transmission of high power laser energy, though there is a problem currently being addressed of a high surface reflection coefficient. It is very probable that CVD diamond-like films will form good anti-reflection coatings for diamond.

In new applications, the technology of making diamond lenses is being developed. The use of diamond as a detector of ionising radiation is well known, but recent work shows its possibilities in thermoluminescent as well as conduction and pulse counting modes. There are further possibilities of using diamond for the detection and measurement of optical radiation. Examples are low, medium, and high intensity far ultraviolet (< 225 nm) and very high intensity near ultraviolet and visible light from excimer, dye, or argon lasers. Diamond is very radiation resistant! Sensitivities, response times and impurity trap levels have been measured and appropriate diamonds can be synthesised. The use of diamond as fast opto-electronic switches has been reported in the literature and the mechanical and thermal design of diamond "heat sink" substrates for semiconductor laser diodes is advancing rapidly.

1. TRANSMISSIVE COMPONENTS

Pure diamond has the widest spectral transmission range of all known solid materials. At the short wavelength end of the optical spectrum it is limited by the absorption due to electronic transitions across the gap between filled and conduction bands. This fundamental absorption edge begins at 5.4 eV or 230 nm wavelength. Above this wavelength the transmission rises fairly rapidly. Figure 1 illustrates this for a pure (type 2A) natural diamond, and also shows the effect of nitrogen impurity in one of the more common type 1 diamonds. The additional absorption centers associated with this nitrogen give a secondary absorption edge which commonly lies in the range 3.7 to 4.5 eV (335-275 nm). [1,2]

Though diamonds exist in a wide variety of colors, ranging from blue to green to yellow to pink and deriving these colors from the absorption systems of impurity centers or radiation-induced damage, substantial quantities in the gem and better industrial grades are white or near white. The commonest non-white colors are yellow and brown. Yellow diamonds derive their color from dispersed nitrogen, whereas the light brown color of many natural type 2A diamonds may well be due to micro-inclusions [3], possibly associated with mosaic boundaries in the crystal structure. White diamonds are of course highly transparent throughout the visible spectrum.

The most noticeable feature of the infrared absorption spectrum is a multiphonon system extending from about 2.5 μm to about 6 μm wavelength. This is due to vibration modes of the C-C bonds in the diamond lattice and is common to all diamonds. However, it is not a very strong system, the absorbance of the strongest peak being about 12 cm^{-1}. This system thus has little effect on the infrared transmission of diamond windows of below say 0.25 mm thickness. Diamond windows of 0.1 or 0.2 mm thickness in diameters to about 5 mm are in general more than adequately strong to withstand a pressure differential of 1 atmosphere, and such windows are thus also usable in the 2.5 to 6 μm range.

Above 7 μm wavelength, pure (type 2A) diamonds show no further absorption systems to 40 μm and beyond. Far infrared transparency of diamond is the subject of experiments currently in progress. The less pure type 1 diamonds show a complex system of absorptions between 7 and 10 μm, due to nitrogen in different states of aggregation. Figure 2 shows one such spectrum as well as the transmission of a type 2A specimen. Different type 1 samples show different relative intensities of the different peaks and the absolute absorbances can be quite high (to 95 cm^{-1} in an extreme case [4]).

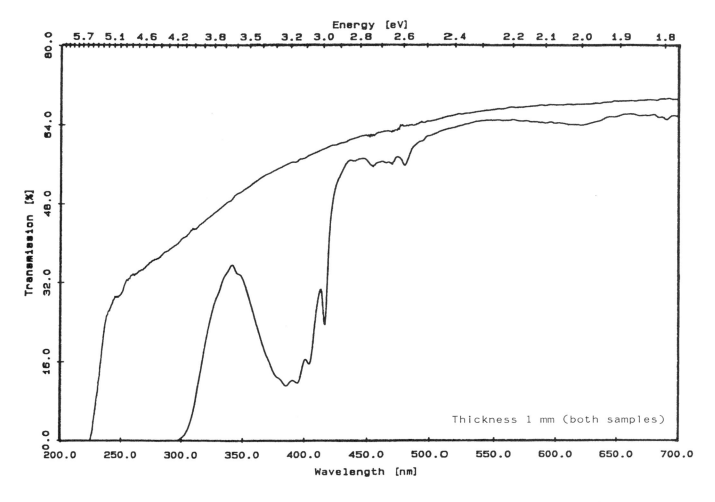

Figure 1. Ultraviolet and visible spectra of a pure type 2A diamond (upper curve) and of a type 1A diamond containing nitrogen impurity (lower curve).

Besides good transmission type 2A diamond has other advantages as an optical material. The first is chemical inertness. Diamond is not attacked by any chemical reagents at the temperatures and pressures commonly employed in chemical laboratories. It can for instance safely be cleaned in reagents such as boiling aqua regia or hydrofluoric acid. Probably the least aggressive materials which will attack diamond are strong oxidising agents such as fused potassium nitrate at temperatures above 450°C. Even with such an active material the etch rate is not high unless the temperature is increased to above 600°C. Fused potassium nitrate at 650 to 700°C is commonly used as an etchant to reveal structure in diamond, 10 minutes being an effective time. Steam and oxygen gas will also attack diamond at temperatures above 600°C. Carbide-forming metals such as iron or nickel will dissolve diamond rapidly at temperatures above their melting points.

High thermal conductivity is also an important advantage of type 2A diamond. At room temperature it is the best heat conductor known, having a thermal conductivity of about 20 watts / cm.K, some 5 times that of copper. [4,5] This, combined with very low absorption at wavelengths near 1 μm and near 10 μm, makes the material a possibility for windows for high intensity Nd-YAG or CO_2 laser beams. As indicated above, the extreme strength of diamond permits the use of very thin windows in the walls of vacuum equipment, thus reducing the residual energy absorption. Further the low expansion coefficient reduces lens distortion of centrally heated diamond windows.

With such advantages it would be surprising if diamond did not have some disadvantages. Technically the most important of these is a high reflection coefficient corresponding to its high refractive index (n). At 1 μm n is about 2.4 and the reflection coefficient at a single interface to air calculated as $(n-1)^2 / (n+1)^2$ is about 17%. There is some disagreement in the literature as to the correct value of n at 10 μm [6,7], but the resulting

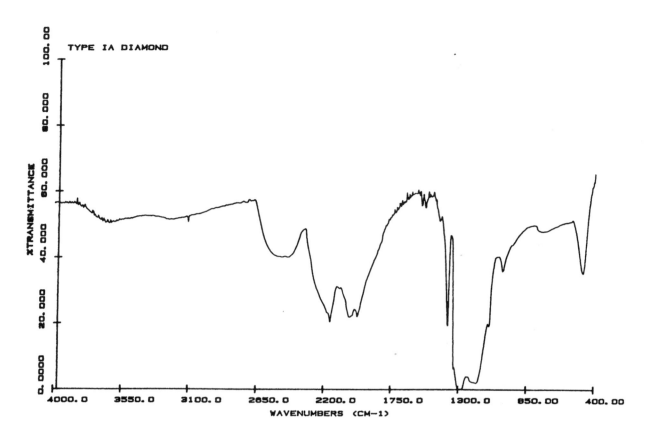

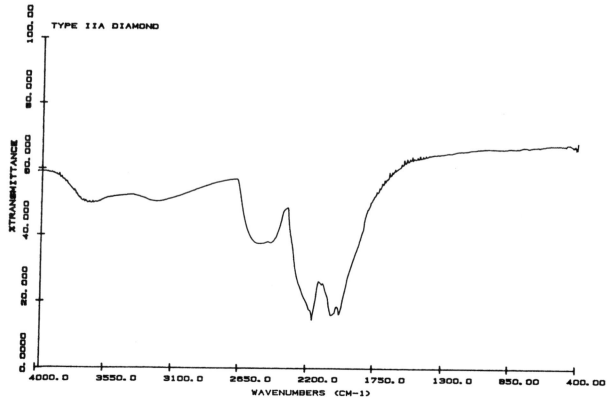

Figure 2. Infrared transmission spectra of type 1A diamond and of type 2A diamond (thickness 1.0 mm in each case).

reflection coefficient will in any case be higher than 15%. An anti-reflection coating would be highly desirable, but it is not easy to find one which does not nullify the advantages of chemical inertness and scratch resistance possessed by the diamond surface. Diamond-like carbon (DLC) coatings are probably the most promising, though their absorbances may be too high. DLC films can be formed by low pressure chemical vapor deposition methods, the most common of which is the cracking of methane-hydrogen mixtures by thermal, microwave, or R.F. energy. This is currently a very active field of research and a recent review [8] and conference proceedings [9] give an entrée to the literature.

Another often stated objection to diamond optics is price. It is true that large diamond windows are expensive, and "large" in diamond terms means anything much above ½ inch (12.7 mm) in diameter. The price is a strong function of size. Typically diamond windows of ½ inch diameter and normal thicknesses are priced in the low five figures in dollars, whilst windows of ¾ inch (19 mm) diameter would be in six figures. There may be problems of availability in these larger sizes and there is an absolute limit set by the maximum size of mined diamonds. A one-inch (25.4 mm) diamond window might or might not be available but it would be extremely difficult to locate and it would almost certainly cost well over a million dollars.

On the other hand, the strong function of price works in favor of the optics designer in the other direction. Small diamond windows are very competitive in price. A 1 mm diameter diamond window typically sells at around thirty to forty dollars and a diamond window 0.5mm diameter by 0.1 mm thick is obtainable for under ten dollars. A thousand dollars currently buys a diamond window a little over 5 mm diameter by 0.25 mm thick.

2. APPLICATIONS

Applications of the larger diamond components are normally in the investigation of difficult environments. Diamond windows have been used for spectroscopic and total radiation flux measurements in space-related applications. Some such components have flown on spacecraft of the Nimbus and Pioneer series and there are others due to fly on the Galileo Jupiter probe. The most spectacular was a diamond window 18.2 mm diameter by 2.8 mm thick in the main pressure shell of the Pioneer Venus large probe launched on August 8, 1978. The diamond was an optical port for an infrared radiometer experiment.[10] Total incoming flux was measured in five spectral bands in the range 3 to 40 μm as defined by separate filters and detectors. Information was obtained as a function of direction by a rotating light-pipe / prism system. The window had to survive Earth atmosphere on launch, the cold and vacuum of space, and the extremes of the Venus atmosphere (about 800 K, 90 atmospheres, largely CO_2 with various acids including H_2SO_4).

Earthbound applications often relate to similarly extreme conditions. Diamond windows are in use in the chemical industry for spectroscopic quality control of molten plastics, caustic alkalis, etc. during processing and pipe transport between processes. Proposals have been made for the use of diamond windows for infrared Doppler shift velocimetry of burning gases in various engines. On a laboratory scale diamond windows have a long history of use in the spectroscopy of fused salts, [11] including such difficult materials as UF_4 in molten LiF - BeF_2 at 550°C. [12] Diamond anvils are used for the generation of ultrahigh pressures (to 4 Mbar and more [13]), but they also act as windows for the observation of the material under high pressure. Reviews of the relevant diamond material parameters [14] and of the technology used in their manufacture [15] (which is also that used in manufacturing other diamond optical components) have been published.

The advantages of diamond include scratch resistance and ease of cleaning. These are factors which influence the choice of materials for instrumentation in the food industry. Diamond optical components have indeed been used in that industry, an example being infrared frustrated total internal reflection prisms for the contact analysis of certain foodstuffs. Cleanliness and scratch resistance of instruments are certainly no less important in surgery, where diamond scalpels and other cutting instruments are used primarily because of their superb cutting edges. [16, 17] Edges can be polished on diamond which are at least 50 times sharper than those of the best razor blades or surgical steel scalpels. Diamond blades are thus used in delicate micro-surgery because they cut cleanly with less force and thus less distortion and displacement of the surrounding tissue and less cell trauma in the wound region. The fact that the blade is transparent means that it can also be used as a light-pipe to bring illumination or laser energy to the cutting edge.

Laser energy at the cutting edge can be used for various purposes. In the first place it seems to assist cutting of some kinds of tissue. Secondly it can be used to cauterise blood vessels, and thirdly it can be used to produce hyperthermia in oncology procedures. Most of these combined cutting-plus-optical instruments are in a very early stage of development, e.g. a knife designed for the laser treatment of "port-wine stain". The problem was to bring laser energy to the enlarged blood vessels beneath the skin surface without burning the overlying skin, and the proposal [18] was to use a flat blade inserted under the skin to reflect and refract the light in a generally downwards direction as shown in Figure 3. From the three-dimensional geometry of the reflection it follows that:[19]

$$\tan \alpha = \frac{1}{2} \tan \theta \sqrt{(\cot^2 \phi + \cot^2 \theta - 1)^2 + 4 \cot^2 \phi}.$$

The light is concentrated towards the center and it turns out that the angles are again quite critical. A nice balance has to be struck between having a blade which is sharp enough to cut and blunt enough to reflect correctly. Current design parameters have ϕ in the range 73 - 75° and θ in the range 40 - 45°.

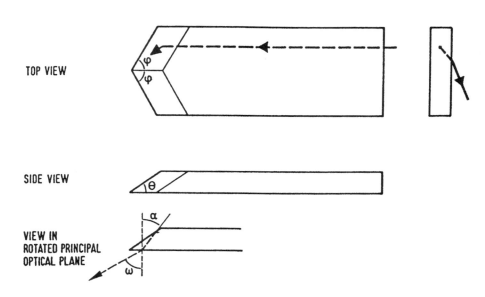

Figure 3. Experimental diamond blade designed to introduce laser energy beneath the skin.

Other uses of diamond in medicine have included various designs of miniature endoscope window, both flat and of prism form, for internal examination via fiber optic bundles. A limitation in infrared imaging technology has been the absence of diamond lenses. These are difficult to produce because the grinding hardness of diamond is a strong function of crystal orientation varying by a factor of perhaps 600 between directions in the crystal. Attainable surface quality is also a function of orientation. The consequence is that attempts to polish spherical surfaces on diamond usually result in figures related to the polar diagram of grinding hardness. Nevertheless some success has been achieved and small diamond lenses are becoming available on an experimental basis.

3. DIAMOND RADIATION DETECTORS

3.1. Introduction

The existence of so-called counting diamonds has been known for at least 40 years. These are diamonds which become conducting under ionising radiation. They can be used in pulse counting or resistive mode and early work was reviewed in 1965.[20] Much of the later work was directed at the measurement of moderately intense gamma radiation used in radiotherapy.[21] Until 1980 it seems that only natural diamond was studied. It was found that there are differences of many orders of magnitude in the responses, due to differences in the impurity and defect structure of the natural crystals. However synthetic diamond of controlled low nitrogen content became available in the early 1980's and this has made possible the development of effective miniature diamond radiation detectors for use in moderate to high dosage gamma, X and other radiation fields. Real time measurements during radiotherapy and in nuclear reactors are applications that come to mind. At the same time thermoluminescent diamond radiation detectors have been developed.[22] These are usable for measurement of cumulative dose at much lower dose rates, possibly in personnel monitoring.

The "other" radiation fields mentioned above include visible, near and far ultraviolet and these are of more direct relevance to the present Conference. Accordingly it seems worthwhile to describe our current experiments in more detail.

3.2. Experimental

Several of the synthetic diamond crystals had been used in previous X-ray investigations.

Others were obtained from Dr. R. Burns (De Beers Diamond Research Laboratory). Cubes of side about 1 mm were cut and polished. Subsequently two opposite faces were sputter coated with a multi-metal layer of 40 nm Ti, 60 nm Pt, and 1000 nm Au to obtain ohmic contacts. The characteristics of the samples are summarised in Table 1 where the nitrogen contents are expressed in terms of the infrared absorption coefficient at 1130 cm^{-1} which is proportional to the concentration of substitutional dispersed nitrogen impurity.[23]

Table 1. Properties of the Diamonds Investigated in our Photoconductance Studies

Specimen Code	Absorption Coefficient at 1130 cm^{-1}	X-ray photo-sensitivity	Photo-current Maximum at 225 nm	Relative Height of $\frac{1}{2}$ s Peak after 20 s off (225 nm)
	(cm^{-1})	(sec/GyΩm)	(A)	
SIIA-4	0.35	2.16×10^{-2}	5×10^{-5}	25%
LN4	0.58	4.69×10^{-3}	10^{-5}	75%
MNI	1.89	5.38×19^{-5}	10^{-8}	95%
S1	5.3	---	2×10^{-11}	10%
S2	5.0	---	5×10^{-10}	45%
S3	9.0	---	10^{-11}	---
S4	9.0	---	5×10^{-12}	---
S5	2.1	---	5×10^{-7}	90%
S6	2.1	---	10^{-8}	95%
Natural 2A	below 0.2	---	2×10^{-10}	---

The metal coated diamonds were clamped between metal contacts and exposed to a focussed monochromatic light beam. To obtain measurable photo-currents the intensity of the light was increased by opening the monochromator slit. The drawback was a low spectral resolution (about 5 nm). The photo-current was measured by a pico-ampere meter interfaced with a personal computer. The bias voltage was kept at 45 V D.C. The spectra were not corrected for the large dependence of intensity on wavelength for the light source-monochromator combination. This correction will be made in future work.

Response times were measured after opening or closing a photographic shutter. The response time of the measuring system was about 0.15 s. The relative influences of shallow and deep traps in the crystals were determined as follows. For shallow traps the crystal was irradiated (225 nm or 375 nm) until no essential increase in photo-current occurred. The shutter was closed and after 20 s reopened for 0.5 s. The height of the photo-current peak as a percentage of the saturation current is a qualitative measure of the concentration of shallow traps. To measure deep traps the crystal was exposed (225 nm) to saturation. The shutter was closed and the crystal exposed to high intensity visible light from a halogen lamp for 30 s. Then the crystal was exposed to UV light again and the photo-current measured as a function of time. The rise time gives a measure of the concentration of deep traps.

3.3 Results: photoconductance spectroscopy

Figure 4 shows typical responses of some low nitrogen diamonds. There is a sharp peak at 225 nm, in some cases accompanied by a shoulder or second peak at about 210 nm. This is intrinsic photoconductance caused by valence to conduction band transitions. There is a strong correlation between the 225 nm peak height and nitrogen concentration (Table 1.). The fundamental optical absorption edge of diamond lies at 225 nm and for shorter wavelengths (and in particular for the 210 nm peak) there is a reduced sensitivity due to the limited penetration of the UV light in the diamond. At 210 nm the photoconductance is a surface effect which samples a much smaller volume of diamond and is thus much more sensitive to the precise impurity content of the material. This effect will play an important role in the development of detectors for low penetration radiation such as alpha particles, very soft X-rays and vacuum UV. Sensitivity is limited by surface charge carrier recombination.

With higher nitrogen the 225 nm peak is reduced to near vanishing point and a broad conductance feature appears in the blue and near UV region (Figure 5). A similar curve is found for natural type 2A samples (Figure 6). Though the sensitivities are low, diamond has a potential advantage for measurements of very intense radiation in that its threshold for radiation-induced damage is very high. The final column of Table 1 shows data related to shallow trap levels in the several samples. Some of the crystals had 90 - 95% re-attainment of peak height in 0.5 s after 20 s off, and thus have low densities of shallow traps. From other work there is evidence that these may be due to B-N pairs in the crystals. Deep traps were found to varying degrees in all the crystals and are the subject of continuing studies.

3.4 Fast optoelectronic switching

There have been a number of reports in the literature of fast optoelectronic switching

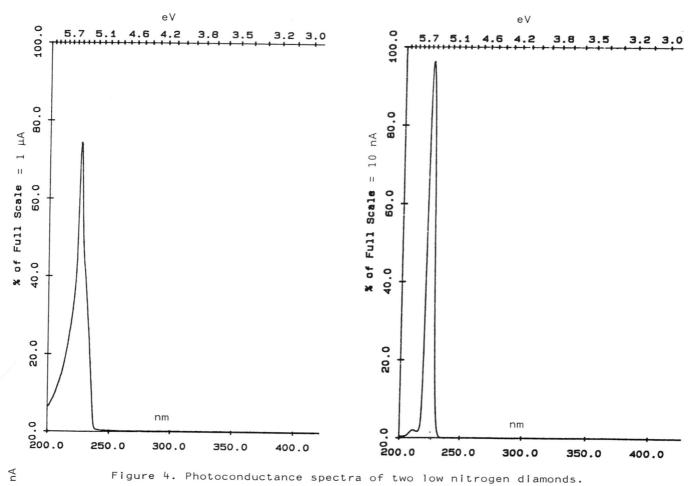

Figure 4. Photoconductance spectra of two low nitrogen diamonds.

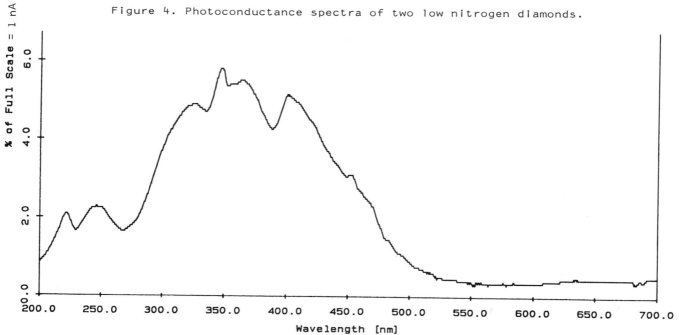

Figure 5. Photoconductance spectrum of a medium nitrogen diamond S1.

in diamond. [24,25] Desirable properties for an optoelectronic switching material include high dark resistivity, high carrier mobility, and high breakdown field strength. Diamond possesses these properties, types 1 and 2A generally having dark resistivities in the range 10^{12} to > 10^{16} Ω cm. Electron and hole mobilities are around 2000 cm^2/Vs and breakdown field around 20 MV/cm. It has been shown [25] that the characteristic response times of six natural diamonds to laser pulse illumination were in the range 125 to 220 ps, with a synthetic diamond showing somewhat faster response (55 ps). The diamonds were under 2 kV DC bias and exposed to sharpened, frequency-multiplied Nd:YAG laser pulses with widths 15 to 30 ps at wavelengths 530, 350, or 265 nm.

Diamond optoelectronic switches with response times in the picosecond range can thus be used for switching high voltage (to several kV) and can be used to generate ultrashort electric pulses or microwave bursts. It seems though that there is a contradiction with the results reported in paragraph 3.3 where times of seconds or minutes were reported as necessary to bring continuously illuminated diamonds to electric saturation, or in other words to fill all the traps. Similar times were needed to empty the shallow traps. Conditions were in fact different in the two sets of experiments with fast, high-intensity, spectrally narrow pulses in the one case, contrasted with continuous, low-intensity, wide-band illumination in the other. Further the one was voltage switching and the other current measurement. Nevertheless the time constants of the trap filling and emptying processes do not seem to be fully understood. Clearly more work, both experimental and theoretical, is needed; and clearly there are interesting devices in the offing.

4. DIAMOND SUBSTRATES

There are of course a whole range of interesting active diamond devices in the offing if the problems of growing epitaxial, controllably doped, single-crystal diamond layers on diamond can be solved. This is the thrust of much of the "New Diamond Technology" based on chemical vapor deposition and ion implantation of carbon on and in diamond. [9] Perhaps it is appropriate to finish this review by referring to the "classical" application of diamond in electronics, that as a heat-sink. Diamond is an extreme electrical insulator and an extreme thermal conductor. That combination is unique at room temperature, a manifestation of the high Debye temperature of diamond and shift of the peak in the curve of thermal conductivity against temperature to higher temperatures. The successful use of diamond as a heat sink for microwave diodes was reported some twenty years ago [26] and in the period since then diamond heat sinks have, amongst other things, made possible the first room temperature semiconductor diode lasers. The thermal design of such lasers can now be optimised by the use of diamond heat sinks of the correct dimensions (which can be calculated [27,28]). These diamond heat sinks are necessarily small, but the aim of much of the present work on diamond growth is to reduce the price of large area diamond sheets to the point where diamond can truly become an economic general purpose substrate for all those integrated circuits and active devices which can benefit from it.

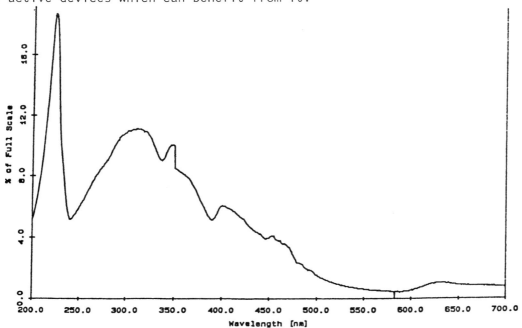

Figure 6. Photoconductance spectrum of a natural type 2A diamond. Full scale on vertical axis = 1 nA.

5. REFERENCES

1. C.D. Clark, R.W. Ditchburn and H.B. Dyer, "The absorption spectra of natural and irradiated diamonds," Proc. Roy. Soc. London A 234, 363-381 (1956).
2. G. Davies and M.H. Nazaré, "Optical study of the secondary absorption edge in type 1a diamonds," Proc. Roy. Soc. London A 365, 75-94 (1979).
3. J.P.F. Sellschop, "Nuclear probe studies," in The Properties of Diamond, J.E. Field ed., Academic Press London, 131-133 (1979).
4. E.A. Burgemeister, "Thermal conductivity of natural diamond between 320 and 450 K," Physica 93B, 165-179 (1978).
5. R. Berman, E.L. Foster, and J.M. Ziman, "The thermal conductivity of dielectric crystals: the effect of isotopes," Proc. Roy. Soc. London A 237, 344-354 (1956).
6. D.F. Edwards and E. Ochoa, "Infrared refractive index of diamond," J. Opt. Soc. Amer. 71, 607-608 (1981).
7. M. Seal, "The increasing applications of diamond as an optical material and in the electronics industry," Industrial Diamond Rev. 38, 130-134 (1978).
8. J.C. Angus and C. Hayman, "Low pressure growth of diamond and diamondlike phases," Science, to be published (1988).
9. Diamond and Diamond-like Materials Synthesis, Materials Research Society, Extended abstracts, symposium D, Spring meeting, Reno Nevada (1988).
10. R.W. Boese, J.B. Pollack and P.M. Silvaggio, "First results from the large probe infrared radiometer experiment," Science 203, 797-800 (1979).
11. G.G. Cocks, J.B. Schroeder and C.M. Schwartz, The Spectroscopy of Fused Salts, Battelle Memorial Institute report no. BMI-1185, Columbus Ohio (1957).
12. L.M. Toth, J.P. Young and G.P. Smith, "Diamond-windowed cell for spectrophotometry of molten fluoride salts," Analyt. Chem. 41, 683-685 (1969).
13. W.C. Moss, J.O. Hallquist, R. Reichlin, K.A. Goettel and S. Martin, "Finite element analysis of the diamond anvil cell: achieving 4.6 M.bar," Appl. Phys. Lett. 48, 1258-1260 (1986).
14. M. Seal, "Diamond anvils," High Temp. High Press. 16, 573-579 (1984).
15. M. Seal, "Diamond anvil technology," in High Pressure Research in Mineral Physics, M.H. Manghnani and Y. Syono, ed., Terra Scientific, Tokyo and Amer. Geophys. Union, Washington DC, 35-40 (1987).
16. D.J. Pierse, "Instrument development: report on International Opthhalmic Microsurgery Group meeting, July 1972," Adv. Ophthalmology 30, 30-39 (1975).
17. S. Herbert, "The diamond knife: rather more than meets the eye," Industrial Diamond Rev. 44, 249-253 (1984).
18. W.J. Hoskin, private communication.
19. D.J. Seal, private communication.
20. F.C. Champion and P.J. Kennedy, "The counting properties of diamonds under ionizing radiations," in Physical Properties of Diamond, R. Berman ed., Oxford University Press, 356-370 (1965).
21. E.A. Burgemeister, "Dosimetry with diamond operating as a resistor," Phys. Med. Biol. 26, 269-275 (1981).
22. R.J. Keddy, T.L. Tam amd R.C. Burns, Phys. Med. Biol. 32, 751 (1987).
23. R.M. Chrenko, H.M. Strong and R.E. Tuff, "Dispersed paramagnetic nitrogen content of large laboratory diamonds," Phil. Mag. 23, 313-318 (1971).
24. P.T. Ho, C.H. Lee, J.C. Stephenson and R.R. Cavanagh, "A diamond opto-electronic switch," Optics Comm. 46, 202-204 (1983).
25. P.S. Panchhi and H.M. van Driel, "Picosecond optoelectronic switching in insulating diamond," IEEE J. Quantum Elec. QE22, 101-107 (1986).
26. C.B. Swan, "Improved performance of silicon avalanche oscillators mounted on diamond heat sinks," Proc. IEEE 55, 1617-1618 (1967).
27. J. Molenaar and G.W.M. Staarink, "The optimal form of diamond heat sinks," Proc. First Europ. Symp. Math. in Industry, Kluwer, Stuttgart, 113-126 (1988).
28. J. Doting and J. Molenaar, "Isotherms in diamond heat sinks, non-linear heat transfer in an excellent heat conductor," Proc. SEMI-THERM, IEEE (1988).

Optical Applications and Improved Deposition Processes for Diamond-Like Carbon Coatings

A H Lettington

Royal Signals and Radar Establishment, St Andrews Road, Malvern, Worcs WR14 3PS, UK

ABSTRACT

This paper describes the development and testing of diamond-like carbon coatings using an RF glow discharge technique. Methods used to nucleate the diamond phase to reduce the growth of graphitic impurities and to improve the deposition process are discussed. Various optical, mechanical and medical applications of the coating are included.

INTRODUCTION

There has been an interest in growing diamond films or layers for very many years. The potential applications range from mechanical components to semiconductor devices. Our own requirement was an antireflection coating for germanium infra red optics that was also abrasion resistant and chemically durable. Existing durable antireflection coatings at the time were based on ZnS-Ge multilayers. These were easily scratched and chemically attacked by dead flies, etc. Diamond has an ideal refractive index match to germanium and also has the required mechanical and chemical properties.

One of our major requirements was to be able to coat uniformly and on a routine commercial basis components which were many feet in diameter. We followed with interest the work of Aisenberg and Chabot[1] who in 1971 described the preparation and properties of "Thin Films of Diamond-Like Carbon". An example of this material on glass showed excellent durability and hardness but its adhesion to glass was poor. It used an ion beam source which we felt could not easily be scaled up to coat large convoluted surfaces.

Professor Spear at Dundee University and Professor Holland at Sussex University, both in the UK, were known to be working on glow discharge methods. Early in 1977 we supplied polished germanium substrates to investigate their processes. Their films were of variable hardness but had good infra red transmission. The adhesion of these coatings to germanium was far better than on glass but still not adequate.

Within RSRE we evaluated a variety of deposition procedures, optimised their growth conditions and developed other processes. These processes were scaled up to large components and by 1978 a significant part of this technology was in full production[2]. Since then we have been investigating improved deposition techniques and infra red transparent carbides[3].

DEPOSITION TECHNIQUES AND PROPERTIES OF DIAMOND-LIKE CARBON FILMS

It is well known that diamond is a metastable form of carbon at ordinary temperatures and pressures, graphite being the more stable form. Diamond crystals are formed directly from graphite at very high pressures and temperatures, but there has been a continuing desire to produce diamonds under ordinary conditions of temperature and pressure. Some of the most successful methods have used a dissociated hydrocarbon gas as the source of carbon in the presence of a high concentration of activated hydrogen. A variety of methods ranging from thermal chemical vapour deposition to RF and microwave plasma assisted processes have been used to achieve the dissociation and activation of the gases and are well documented in the literature. In all these methods the presence of the activated hydrogen at elevated substrate temperatures reacts with the graphite that might otherwise be deposited and preferentially grows the metastable diamond form once it has been nucleated.

Our own approach has been somewhat different. It has involved the dissociation of a hydrocarbon gas in an RF plasma. The component to be coated is usually biassed negatively either through the application of a DC potential or else through the diode action of a capacitively coupled RF system. The positive carbon ions are attracted to it producing the deposited coating. The concentration of activated hydrogen is far lower than that described above. Instead we have adjusted the Kinetic Energy of the incident carbon ions to back sputter the more stable graphite which has a lower bond energy compared to diamond so leaving the growing diamond-like layer.

This approach has certain advantages and disadvantages. It has allowed us to coat large convoluted surfaces uniformly. However the high impact energy has tended to build stresses in the coating limiting its maximum thickness to about 2-3 μm. This high impact energy which can be of the order of 0.5-2 KeV, depending on the exact glow discharge deposition method being used[4,5,6], limits the size of the diamond crystallites which are imbedded in a matrix of hydrogenated heavily cross-linked polymer containing graphitic impurities. A high energy electron diffraction study by Loughborough University of a thin film coating supplied by OCLI is shown in Figure 1. From this they have identified the presence of these small diamond crystallites.

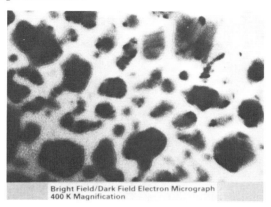
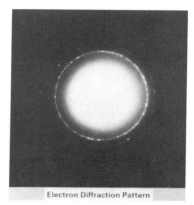

Bright Field/Dark Field Electron Micrograph
400 K Magnification

Electron Diffraction Pattern

Figure 1. Electron micrograph and diffraction pattern for DLC coating.

In the present glow discharge method of deposition, with its continued back sputtering action it is difficult to nucleate the diamond-like carbon except on substrates of silicon, germanium or silicon containing compounds. We have developed a range of intermediate bonding layers that enable the coating to be deposited on most substrate materials.

The properties of diamond-like layers using an RF glow discharge technique have been described in the recent literature[7]. In early unpublished work we explored the range of hydrogenated carbon deposition from soft cross-linked plastics to hard black graphitic layers. The plastic nature occurred at low potentials (~ 100 V) and temperatures while harder coatings were achieved with potentials nearer 1 KV.

The infra red transmission of these layers is strongly dependent on the deposition conditions, particularly the temperature of the substrate. Increasing the temperature drives out much of the hydrogen in the matrix, reducing the spectral absorption lines but at the same time it increases the scattering of the infra red radiation by increasing the graphitic inclusions. A compromise has to be reached between these two absorption processes so as to maximise the infra red transmission while retaining adequate hardness and durability.

This coating has been used on a range of infra red optical components and for other optical, mechanical and medical applications as described in a later section.

IMPROVED DEPOSITION TECHNIQUES

We have followed with interest the current programmes to produce a purer form of synthetic diamond than is available from the conventional glow discharge method and are currently funding parallel activities in UK industry based mainly on microwave techniques. In addition within RSRE we have studied a modified RF glow discharge technique in which excess hydrogen is added to the hydrocarbon feedstock. The effect of the excess hydrogen is to increase the efficiency of the process, and although the main flow of hydrocarbon has been reduced the deposition rate of the coating has remained unchanged at about 1 μm per hour. In our improved deposition technique we have increased the temperature of the substrate in order to compensate for the otherwise increased hydrogen content in our films. The resulting coatings have shown a reduced graphitic scattering while still retaining the excellent IR transmission and durability. These coatings are still highly disordered which we believe is an advantage since it has suppressed the otherwise dominant phonon absorption lines. In pure diamond an absorption at about 10 μm in the infra red region of the spectrum is forbidden due to symmetry. The slightest impurity breaks this symmetry in a diamond crystal resulting in strong infra red absorption. This absorption appears to be present in all diamond layers that we have examined produced by the microwave plasma deposition technique. The effect is masked to a certain extent by SiC absorption bands in the same spectral region. It does however cast considerable doubt as to whether crystalline diamond can be made sufficiently pure and free from gaseous impurities to make a viable bulk window for the 8-14 μm infra red spectral region. We have been investigating alternative compounds for this application.

Diamond layers can still have widespread electrical and mechanical uses and we believe the most promising deposition technique is one which combines the chemical etching due to activated hydrogen with the back sputtering process which is present with high energy impacting ions. These are probably combined in a plasma spraying process which also has enhanced deposition rates.

PROPERTIES OF DIAMOND-LIKE CARBON COATINGS ON GERMANIUM

Initially 25 mm diameter test samples of germanium were coated and subjected to laboratory tests. Then larger components were subjected to various field trials before the process was used to coat windows and lenses for project applications.

The testing was to RSRE Specification TS 1888 and is described in greater detail elsewhere[2]. For the laboratory testing of the abrasion and wear resistance of coatings we developed the abrasion tester illustrated in Figure 2.

Figure 2. RSRE Abrasion Tester.   Figure 3. Windscreen Wiping Test.

The test sample is placed at the bottom of a cylindrical container and must survive being scoured under load in a slurry of sand and water at a 1000 rpm for 5 minutes. The equivalent field trial involved a simulated windscreen wiper mounted on the outside of a vehicle undergoing extensive driving trials, see Figure 3.

Sea trials involved a salt water immersion test. In the winter of 1978 two of our coated discs of germanium were mounted at sea level on a fort in the Solent along with an uncoated square of germanium. They remained at sea continuously for 4½ months. The effect of the sea water at the end of the trial can be seen in Figure 4.

Figure 4. Results of Salt   Figure 5. Coated Window   Figure 6. Coated Window
Water Immersion.   for Helicopter Pilot Aid.   on Aircraft Nose Cone.

The coated samples were virtually unaffected while the uncoated sample was badly corroded and etched away. Examples of the use of diamond-like carbon antireflection coatings are shown in Figures 5 and 6. A typical optical performance of the now commercially available coating is illustrated in Figure 7.

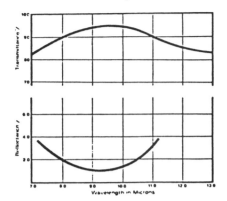

Figure 7. Typical Performance of a Commercially Available Hard Carbon Coating.

This spectral graph shows transmittance of a 1mm thick germanium substrate coated on the front surface with Hard Carbon coating to OCLI specification 6040011 with a High Efficiency Antireflection coating to OCLI specification 6040008 on the rear surface with the following performance:

T ≈ 90% average 8-11 microns Peak T. of 95% (Typical)
R ≤ 3.5% average 8-11 microns Minimum R. of 1% (Typical)

APPLICATIONS OF CARBON COATINGS TO FRONT SURFACE ALUMINIUM MIRRORS

In current thermal imaging systems front surface mirrors produced by single point diamond machining of bulk aluminium are used for rotating polygons, flapping mirrors and for relay elements with optical power. The optical performance of these components tend to deteriorate with time and with exposure to the atmosphere. This process can be prevented through the use of a suitable optical coating. Unfortunately the reflectivity of these coated surfaces can be low when used at oblique incidence. This effect has been demonstrated[8,9] in aluminium mirrors protected with thin overcoatings of SiO_x and intended for use in the 8-12 µm spectral band. This effect occurs for only one direction of polarisation, R_p, parallel to the plane of incidence. Similar effects are observed for many other protective coatings and other metallic reflectors[9,10] making these coatings unsuitable for use in 8-12 µm thermal imaging systems on 45 degree mirrors or scanning polygons.

We[11] identified the origin of this effect and demonstrated that it does not occur in diamond-like carbon protective coatings on these front surface mirrors. The problem with coatings such as SiO_x is that they have strong optical absorption lines in the spectral region of interest. The coatings are sufficiently thin for this to produce negligible absorption. It does however affect the values of n and k in the absorbing region such that the Brewster angle at the air-coating interface occurs at very low angles of incidence. There is destructive interference between this reflection and that at the coating-metal boundary resulting in a loss of reflectivity. This effect does not occur with diamond-like carbon coatings on these mirror surfaces. An example of an RSRE coated open catadioptric infra red telescope is illustrated in Figure 8. Similar coatings are now available commercially.

Figure 8. RSRE Coated Catadioptric Telescope (Courtesy Rank Pullin Controls).

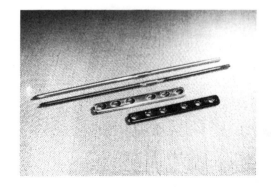

Figure 9. Carbon Coated Orthopaedic Pins.

THE USE OF DLC COATINGS FOR PHOTOTHERMAL CONVERSION OF SOLAR ENERGY

The main aim of photothermal solar energy conversion is to collect solar radiation and to convert it into useful heat. There are two main types of converter: the flat plate collector, where an area of an absorbing material is placed so as to collect the solar radiation; and the focussing collector, where solar radiation is condensed onto a smaller absorbing area. It is usual to remove the heat from both systems with circulating water. For either collection processes to be effective there must be a maximum absorption α of the solar radiation and minimum heat losses in particular a minimum thermal emittance ε at infra-red wavelengths.

The surface must have high absorption (low reflectivity) from 0.3 to about 1.7 µm and low emission (high reflectivity) above 2 µm, with a sharp transition between these two regions.

Several spectrally selective coatings have already been proposed and made using silicon or germanium layers deposited onto polished substrates[12-15], and some time ago we[16] proposed the use of diamond-like carbon for this purpose. We measured the absorption coefficient of our diamond-like carbon coatings over the visible and infrared regions (see Figure 10) and from this and the measured refractive index of about 2.2 calculated the values of α, ε and α/ε for C, Si and Ge single layer coatings of varying thickness deposited onto aluminium (see Table 1).

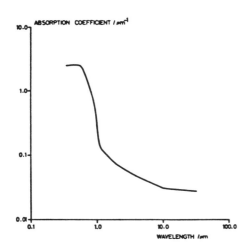

Figure 10. Absorption Coefficient of Carbon.

Thickness (um)	C			Si			Ge		
	α	ε	α/ε	α	ε	α/ε	α	ε	α/ε
0.5	0.70	0.05	14.0	0.53	0.07	7.6	0.50	0.10	5.0
1.0	0.77	0.17	4.5	0.56	0.07	8.0	0.52	0.08	6.5
1.5	0.79	0.23	3.4	0.54	0.06	9.0	0.52	0.07	7.4

Table 1. Solar Absorptance and Thermal Emittance of Single Layer C, Ge and Si Coatings on Aluminium.

The optical properties for Al and Si were taken from Ref 17 and those for Ge from Ref 18. From Table 1 it can be seen that a single layer of carbon has the highest efficiency. This value however is not high enough for most applications and we have sought ways to improve the α/ε ratio in multilayer designs. Our best multilayer coating (α/ε = 19) employed a 0.09 µm thick layer of low index material (n = 1.45) on top of a 0.4 µm thick layer of carbon.

MECHANICAL APPLICATIONS OF DIAMOND-LIKE CARBON LAYERS

Apart from the infra red properties of diamond-like carbon, discussed in the previous section, the material is also hard and chemically durable and is useful for protecting metal objects from scratching and chemical attack. We have coated a variety of metal objects ranging from large sheets to nails, twist drills and machine tool inserts. Some of these have remained in the open exposed to the atmosphere for the last seven years without deterioration. The machine tool insert was used for the high speed cutting of aluminium and lasted longer than uncoated inserts. In these cases the commercial viability of the coating process was very finely balanced.

An area in which we had greater success was in coating thin optical fibres. A carbon coating inhibited the attack of the silica fibre by moisture so preventing a site for subsequent brittle fracture from occurring.

MEDICAL APPLICATIONS OF DIAMOND-LIKE CARBON COATINGS

One application that we investigated was coating orthopaedic pins (see Figure 9). Where they come in to contact with the air is frequently a place where infection takes place. In an attempt to overcome this problem we coated a set of pins which were used in sheep trials by Professor McGibben of Cardiff Royal Infirmary. The trials were successful and the tissue knitted to the coated pins. Far more laboratory testing needs to be done before they could be accepted in human trials, but if acceptable one could imagine coating many other implants such as the roots of false teeth.

We are probably just at the beginning of an exciting new era in the applications of diamond-like carbon coatings.

REFERENCES

1. Aisenberg, Chabot R, J Appl Phys, Vol 42, p2953, 1971.
2. Lettington A H, SPIE Conference Proceedings, Vol 590, pp100-105, Cannes, 1985.
3. Lettington A H, Lewis J C, Wort C H J, Monachan B C, Hope A J N, E-MRS Meeting, Vol XVII pp 469-474, 1987.
4. Anderson D A, Phil Mag, Vol 35, pp 17-26, 1977.
5. Ojha S M, Holland L, Thin Solid Films, Vol 38, L17, 1976.
6. Whitmell D S, Williamson K, Thin Solid Films, Vol 35, pp 255-261, 1976.
7. Dischler B, Bubenzer A, Koidl P, Sah R E, Spring Conference on Applied Optics, Paper TuA-D3, Monterey, 1984.
8. Cox J T, Hass G, Hunter W R, Appl Opt, Vol 14, p1247, 1975.
9. Pellicori S F, Appl Opt, Vol 17, p3335, 1978.
10. Cox J T, Hass, G, Appl Opt Vol 17, p333, 1978.
11. Lettington A H, Ball G J, RSRE Memorandum No 3295, 1981.
12. Hahn R E, Seraphin B O, Physics of Thin Films, Vol 10, p1, 1978.
13. Drummeter L F, Hass G, Physics of Thin Films, Vol 2, p305, 1964.
14. Seraphin B O, Thin Solid Films, Vol 57, p293, 1979.
15. Janai M A et al, Solar Energy Materials, Vol 1, Nos 1&2, p11, 1979.
16. Ball G J, Lettington A H, RSRE Memorandum No 3617, 1983.
17. Gray D E (Ed), American Institute of Physics Handbook, McGraw Hill, New York, 1972.
18. Brattain W H, Briggs H B, "Optical Constants of Germanium", Bell Telephone Systems Technical Publications, Monograph B - 1979.

SURFACE MORPHOLOGY AND DEFECT STRUCTURES IN MICROWAVE CVD DIAMOND FILMS

Koji Kobashi, Kozo Nishimura, Koichi Miyata, Yoshio Kawate
Electronics Technology Center, Kobe Steel, Ltd.,
1-5-5, Takatsukadai, Nishi-ku, Kobe, Japan 673-02
and
Jeffrey T. Glass, Bradley E. Williams
Department of Materials Science and Engineering,
North Carolina State University, Box 7907, Raleigh, NC 27695-7907

ABSTRACT

Polycrystalline diamond films were deposited by the microwave-plasma chemical-vapor-deposition (CVD) on Si substrates using a mixture of methane and hydrogen for the source gas. In the morphology study of diamond films using a scanning electron microscope (SEM), it was found that upon increasing the methane concentration (hereafter denoted by c in units of vol%), the surface texture changed discontinuously from (111) to (100) at around c=0.4%, and gradually from (100) to microcrystalline above c=1.2%. The diamond-Si interfaces and the defect structures in the films were investigated by transmission electron microscopy (TEM). The film growth process was investigated by SEM, and it was found that the appearance of small grains and the formation of well-defined diamond faces took place repeatedly with time during the CVD synthesis. The film morphology of boron-doped diamond films on Si substrates and on non-doped diamond films were also presented.

1. INTRODUCTION

Diamond films synthesized by CVD consist of fairly large crystalline grains and exhibit a variety of crystal habits as a function of c. This feature allows us to study the morphology and the defect and interface structures by SEM and TEM, respectively, for the purpose of obtaining basic knowledge which will aid the growth of defect-free epitaxial diamond films for electronic applications. Presented in this paper are the SEM and TEM data of non-doped diamond films, and SEM data of B-doped films. An emphasis is placed upon the infrared (IR) absorption of these films.

2. EXPERIMENT

The films were deposited on Si substrates by microwave (2.45 GHz)-plasma CVD using a methane-hydrogen mixed gas. The experimental set-up, schematically depicted in Fig. 1, is similar to that established by Kamo et al.[1] The gas pressure used was 30 Torr and the flow rate was 100 cc per minute. The substrate, 2 cm x 1 cm in area and 0.5 mm thick, was immersed entirely in the plasma during the CVD. The substrate temperature was measured by an optical thermometer and kept at 800 C by adjusting the microwave power to about 300 W.

In order to increase the nucleation density of diamonds, Si wafers were polished with a diamond paste of 1/4 micron size for one hour, cut to size, washed by distilled water, alcohol and acetone using an ultrasonic cleaner, and finally dried in the air. Using such substrates polycrystalline diamond films were grown at a growth rate of about 0.2-0.3 micron/h.

3. MORPHOLOGY OF THE FILM SURFACES[2]

Figure 2 shows a side view of a film deposited for 7 h on a Si(100) substrate for c=1.0%. It is clearly seen that the film underwent a columnar growth. Figure 3 schematically summarized the morphology of the film surfaces. At c=0.2%, only particles were grown, which have various cubo-octahedral shapes. At c=0.3%, the substrate was thinly covered with diamond particles having triangular (111) crystallographic planes. Most particles were not single crystals but have twin structures. The surface morphology changes markedly at c=0.4 %, where the majority of the crystallographic planes appearing on the film surface were (100), characteristic of the planar square surfaces, overlapped by the secondary growth of diamonds. The (100) features become more prominent as c increases to 1.0%. Note that the sides of the (100) face are very rough, suggesting that (100) faces underwent an anisotropic growth, almost perpendicularly to the Si substrate plane. As the concentration increases beyond c=1.2%, the density of (100) faces decreased and the square feature gradually faded. For concentrations higher than c≃1.6%, the film surfaces become totally microcrystalline and no crystallographic planes of diamond are observed. The fact that the surface morphology changes from (111) to (100) in a very narrow range of the methane concentration around c=0.4% indicates how critically the morphology depends on c, and presumably also on other thermodynamic conditions of the plasma. This is in strong contrast to the fact that the microcrystallization of the film takes place over a relatively wide concentration range above 1.2%.

X-ray and RHEED spectra of the films showed that the films were composed of crystalline diamond even for the highest methane concentration of 2.0 %. On the other hand, the Raman data showed that more graphitic and amorphous carbon were present in the films as c increased.

IR absorption of the films was measured between 400 cm^{-1} and 4000 cm^{-1}, and the absorption was found only in the C-H vibrational region around 3000 cm^{-1}. However, as seen in Fig. 4a, the absorption intensity of the films synthesized for 7 h are so weak that no systematic trend with c was found. The spectra of thicker films, synthesized for 97 h at c=0.3% and for 52 h at c=1.2%, shown in Figs. 4b and c, respectively, have a main peak at 2920 cm^{-1} and a broad band around 2850 cm^{-1}. The former arises from CH and CH_2 vibrations where the C atoms have sp3 bonds with adjacent C atoms, whereas the latter arises from CH_2 vibrations where the C atoms also have sp^3 bonds. A rough estimate showed that the specimens of Fig. 4b and c contain only 3 to 4 at% of hydrogen. The hydrogen content in the diamond films is much smaller than in a-C:H films, which usually contain more than 10 percent of hydrogen, because the substrate temperature for the diamond deposition was maintained at 800 C during the synthesis. It is argued that the hydrogen is chemisorbed only on the film surface and at the grain boundaries, and not contained in the diamond grains.

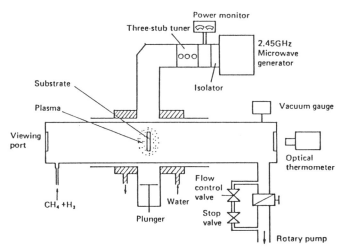

Figure 1. Microwave Plasma CVD Apparatus.

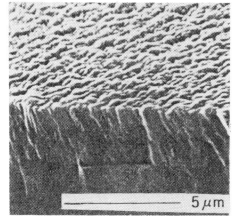

Figure 2. Fractured edge of a diamond film.

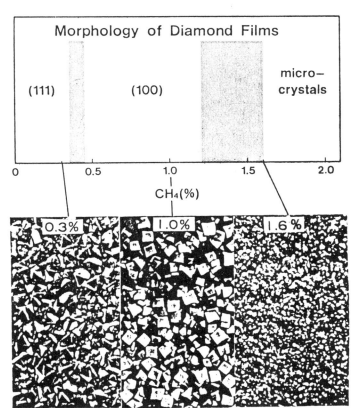

Figure 3. Morphology of diamond films.

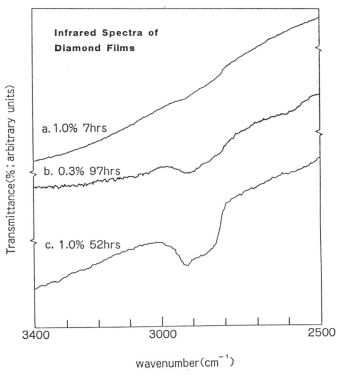

Figure 4. IR spectra of diamond films.

4. INTERFACE AND DEFECT STRUCTURES

Figure 5 and Figs. 6a and 6b show high-resolution cross-sectional TEM micrographs of the diamond-silicon interfaces of the specimens grown at c=0.3% and 2.0%, respectively. For the case of c=0.3% (Fig. 5), an interfacial layer of beta-SiC was observed. It is noted that {111} cross fringes observed in the SiC layer are aligned epitaxially with the fringes observed on the Si substrate. The thickness of the SiC layer is approximately 50 A. For the case of c=2% (Fig. 6), no SiC layer was present. The diamond grain observed in Fig. 6a is epitaxially twined relative to the Si substrate. However, not all the diamond grains showed this twinned relationship as illustrated in Fig. 6b where two diamond grains are misoriented relative to the substrate. The widely spaced fringes in the center of Fig. 6b are Moire rings caused by the overlap of the two diamond grains. Other TEM data showed that diamond grains contain a high density of defects but the defects concentration is reduced at lower methane concentrations.

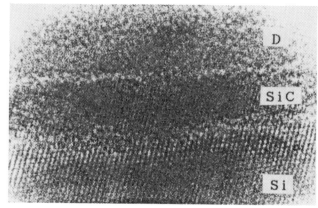

Figure 5. High-resolution cross-sectional TEM: c=0.3%.

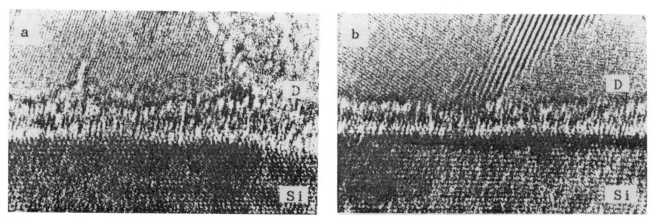

Figure 6. High-resolution cross-sectional TEM: c=2.0%.

5. GROWTH OF DIAMOND FILMS[2]

In order to observe the evolution of surface morphology of diamond films, the CVD synthesis was interrupted every few hours and the specimen was taken out of the reaction chamber to examine exactly the same position on the film surface by SEM. Among several runs undertaken for different experimental conditions, only the major results of c=1.2% are presented in Figs. 7a-7d. After 3 h, the substrate was already covered with small grains which have square (100) faces. After 8.3 h, a secondary growth took place and the (100) faces were being covered with small crystallites. After 11.3 h, the small crystallites disappeared almost completely and the surface was covered with (100) faces of different sizes. This feature persisted through 17.3 h to 26.3 h (Fig. 7a), and a tertiary growth began in between the (100) faces. Again the film surface was gradually covered with small crystallites (Fig. 7b and 7c). However, the tertiary crystallites disappeared completely in Fig. 8d, and the film surface consisted of well-defined (100) features only. The synthesis was continued up to 60 h and the cyclic behavior of the higher-order growths followed by the conversion to the (100) structure was confirmed. A similar cyclic behavior was also observed in the growth morphology for c=0.3% and 0.8%, but the changes were less clear, partly because the periods of this growth cycle are longer than for c=1.2%.

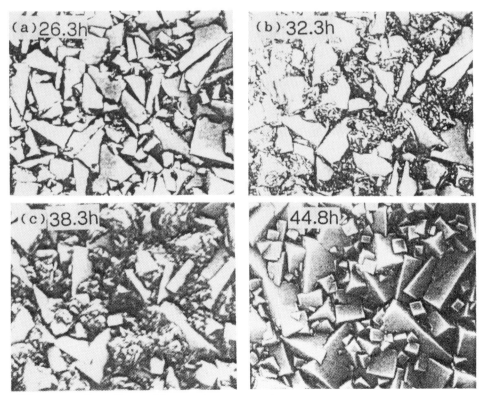

Figure 7. Growth process of a diamond film synthesized using c=1.2%.

6. MORPHOLOGY OF B-DOPED FILMS

Boron-doped, p-type semiconductive diamond films can be synthesized by mixing diborane (B_2H_6) in the methane-hydrogen source gas. Figs. 8a-8c show the SEM photographs of the film surfaces of non-doped and B-doped films grown on Si substrates with c=0.5%, where the diborane concentrations (c_d) are indicated. As seen in Fig. 9a, the non-doped film surface consisted of randomly-oriented, triangular (111) faces. The surface of the B-doped film with c_d=1 ppm (the B to C ratio in the reaction gas was B/C=0.04%) also consisted of (111) faces but each grain had a pyramidal shape as seen in Fig. 9b. For the film of c_d=50 ppm (B/C=0.2%), particles with needle-like structure on the surfaces were deposited. X-ray analysis of the deposit showed that the particles consisted of graphite and no diamond was included. In order to investigate the influence of the substrate material on the morphology of B-doped deposits, B-doped films were deposited on non-doped diamond films grown for c=0.5% using a diborane-mixed methane (0.5%)-hydrogen gas. The SEM photographs of a non-doped diamond film, B-doped (c_d=1 ppm, B/C=0.04%) diamond film, and B-doped (c_d=50 ppm, B/C=0.2%) deposit are shown in Figs. 9a, 9b and 9c, respectively. Like the previous case, the pyramidal grains were grown for c_d=1 ppm, and the needle-like structure is observed for c_d=50 ppm. Notice that in the latter case, the needle-like deposit exists only on the triangular (111) faces and the square (100) faces have no such deposit. Similar experiments were repeated using the non-doped films grown with c=1.2 %, whose surface consists of (100) faces of diamond grains as seen in Fig. 10a. SEM photographs of B-doped films grown using c=1.2% with c_d=1 ppm (B/C=0.09%) and with c_d=50 ppm (B/C=0.48%) are shown in Figs. 10b and 10c, respectively. As seen in Fig. 10b, like the previous two cases, the pyramidal grains were grown on the non-doped film despite the fact that the non-doped film consisted of (100) diamond faces. For the case of c_d=50 ppm, the entire film surface was covered by the needle-like structure as seen in Fig. 10c, and no selective deposition was seen. These data shows that the morphology of B-doped deposits is not significantly influenced by the morphology of the substrate.

IR absorption spectra of the B-doped films deposited on Si substrates and on the non-doped diamond films grown with c=0.5% and 1.2% with different c_d are shown in Figs. 11-13, respectively. It is clearly seen that the absorption is enhanced rapidly with the diborane concentration c_d.

7. CONCLUSION

The surface structures of diamond and B-doped films was studied by SEM, and it was found that the morphology is sensitively dependent on the methane and diborane concentrations. The study of film growth showed the periodic change of surface morphology during the CVD process. The TEM study of non-doped diamond films revealed the diamond-substrate interface structures.

Figure 8. B-doped films deposited on Si: c=0.5%.

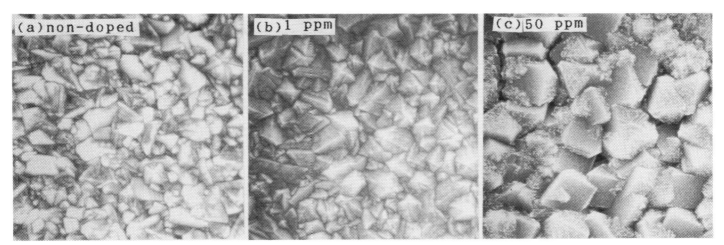

Figure 9. B-doped films (c=0.5%) overgrown on non-doped
diamond films (c=0.5%).

Figure 10. B-doped films (c=1.2%) overgrown on non-doped
diamond.films (c=1.2%).

———————— 3 μm

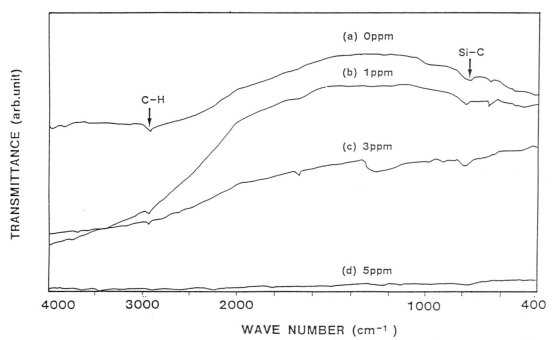

Figure 11. IR spectra of B-doped films deposited on Si

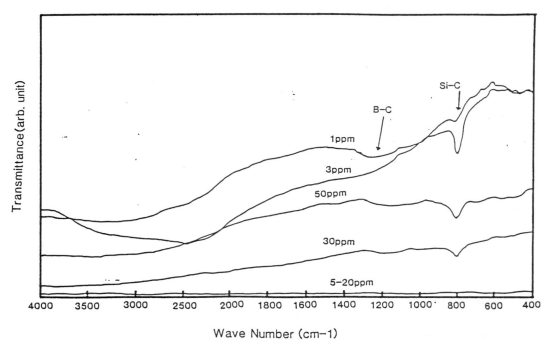

Figure 12. IR spectra of B-doped films (c=0.5%) overgrown on non-doped diamond films (c=0.5%).

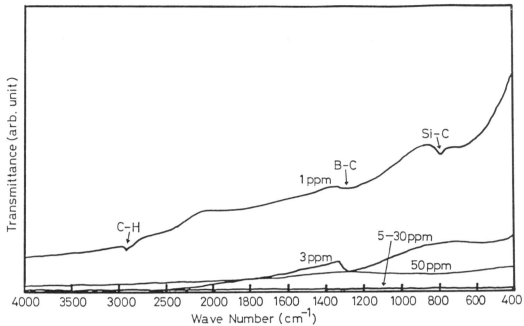

Figure 13. IR spectra of B-doped films (c=1.2%) overgrown on non-
doped diamond films (c=1.2%).

8. ACKNOWLEDGEMENT

The authors wish to acknowledge Prof. R. Davis of NCSU for useful discussion and encouragement for the NCSU-KSL research collaboration on diamond films. Ms. K. More and Dr. J. Posthill are also acknowledged for their assistance.

9. REFERENCES

1. M. Kamo, Y. Sato, S. Matsumoto and N. Setaka, "Diamond Synthesis from Gas Phase in Microwave Plasma," J. Cryst. Growth, 62, 642-644 (1983).
2. K. Kobashi, K. Nishimura, Y. Kawate and T. Horiuchi, "Synthesis of Diamonds by Use of Microwave-Plasma Chemical-Vapor-Deposition: Morphology and Growth of Diamond Films," Phys. Rev. August (1988).

THE POTENTIAL OF DIAMOND
AS A VERY HIGH AVERAGE POWER TRANSMITTING OPTICAL MATERIAL[*]

Sidney Singer, DRA/LT
Los Alamos National Laboratory
Los Alamos, NM 87545

1. Introduction

Optics are an integral part of lasers, whose size and power output have grown greatly in the last several decades. As the power levels grow, the need for compatible high-power optical materials has also grown. Laser designers know that the limits on intra-cavity and extracted power are set by the damage to the laser optics rather than by what can be generated in the laser medium. The limits on laser power imposed by materials properties and their processing apply to high peak and average power systems and to reflecting and transmitting optics.

2. Limitations of Traditional Materials

Some of these limits can be appreciated by examining Figure 1, which shows a typical edge-cooled AR-coated glass optic (for simplicity, the AR coating is shown on only one surface). The thin film AR coating layers often have an absorption coefficient for light that is many orders of magnitude greater than the glass substrate. Even though the films are thin, the production of heat coupled with the poor thermal conductivity of the films and glass substrate can lead to early failure of the coating. The presence of absorbing inclusions exacerbate such problems and additionally may be a preferred site for electron avalanching and coating failure at high peak power. The possibility of foreign material at the surface (e. g., by-products of the grinding-polishing process) are another source of production of heat without a satisfactory method of getting rid of it. And, on the surfaces of the optic, there are inevitable absorbing surface contaminants whose composition depend on the external environment and aren't necessarily constant in time.

For transmission of high power optical beams, glass itself is far from an optimum material. Although (very pure) glass can be made to have a small bulk absorption coefficient for light, its thermal conductivity is so low that the removal of heat produced internally (from bulk absorption or absorbing inclusions) is limited. As a result, high incident power can produce temperature gradients in the glass that lead to degradation of the beam quality and the eventual failure of the optic and its coating system.

In an idealistic limit, one might suppose that all of the coating issues could be resolved and one were left with the properties of glass itself. What then are the limits on power transmission?

In Figure 2, we show the results of a simple calculation of heat transmission in a circular glass slab. The optic is assumed to be 1 cm in diameter and the incident light beam is 1 mm in diameter. The surface is assumed to be insulated and surface absorption effects (i. e., coatings, etc.) are ignored. The edge of the slab is held at a fixed temperature. In the steady state, the temperature distribution in the slab is described by the Poisson equation

$$\nabla^2 T(r) = - \frac{I_0 \alpha}{K} \qquad (1)$$

where T(r) is the temperature, α is the optical absorption coefficient [cm^{-1}], K is the thermal conductivity [watts/(cm-deg Kelvin)], and I_0 is the incident optical intensity [$watt/cm^2$]. Figure 2 shows how the temperature difference between the center of the slab and a point at radius r varies with r. Inside the beam, the radial dependence is parabolic; outside it is logarithmic. The constraint on incident power derives from the temperature difference across the region where the beam transits the optic; in the context of this simple model, that temperature difference is

$$T = (P_0/4\pi) \ (\alpha/K) \qquad (2)$$

The limits arise because the optical path length through the optic depends on temperature. The optical index n itself depends on temperature, and the path length L through the material also depends on temperature because of the material expansion with temperature. There are also effects due to the gradient itself (photoelastic effects) but we ignore them for the sake of simplicity. In evaluating the effects of optical path difference, it is traditional to assume that the permissible difference is about 1/10 of a wavelength λ. This condition is expressed by

$$d(nL) = nL \ (dL/L + dn/n) \quad T = \lambda/10 \qquad (3)$$

For glass, n ~ 1.5; $dL/L + dn/n$ ~ 5×10^{-6}; and we assume L = 0.5 cm. We conclude that the temperature difference between the center and the edge of the light beam cannot exceed a few degrees. In this simple model, the important parameter is K/α.

The implication of this result is shown in figure 3. Using (2), we have plotted the maximum incident _power_ as a function of the parameter K/α for temperature differences of 1 $^\circ$K and 10 $^\circ$K. Superimposed on this plot are the ranges in K/α for glass and a variety of other materials. We conclude that glass at the size assumed can at best transmit 1 - 50 KW (although single-crystal quartz at 12 $^\circ$K can transmit 1 - 10 MW). But there are several materials -- LiF, sapphire, and diamond -- that seem to be able to transmit powers of 10^8 - 10^{10} watts. To facilitate comparison among these materials, their properties [1] have been listed in Table 1. The column labeled T_{opt} shows the temperature at which the thermal conductivity has its maximum value, and the label "cold" in figure 3 implies that the indicated material is assumed to be at that temperature.

Thus, while the power transmission through _glass_ is limited, there are a number of crystalline materials whose properties seem to be compatible with the transmission of very high powers. This implied capability arises mainly because of their high thermal conductivity, and among them diamond is pre-eminent; it may be the most useful because of the accessibility of the temperature required to obtain high thermal conductivity. What is the physics behind electrically insulating materials that possess such high thermal conductivities?

3. Thermal Conductivity in Nearly Perfect Crystals [2]

For "perfect" crystals, the thermal conductivity approaches infinity. In such materials, energy propagates via lattice vibrations (phonons). Here "perfect" means a perfect lattice (a single crystal free of dislocations, internal boundaries, etc.), perfect purity (no chemical or isotopic impurities anywhere in the lattice), and a perfectly harmonic (linear) force relationship between the atoms at the lattice sites. Under such conditions, phonons can propagate through the crystal without loss and heat can thus be transported in the crystal with no temperature gradient.

Real materials fall short of these idealizations and the thermal conductivity is finite. The forces between atoms are not linear; this gives rise to Umklapp (or "U") processes that involve phonon-phonon scattering. Most crystals have defects -- chemical and isotopic impurities and lattice imperfections -- and all materials are of finite size. All of these effects impede the propagation of phonons and cause a finite thermal conductivity. The effect on K is temperature-dependent and is shown in figure 4, where the the thermal conductivity of a generic material is shown as a function of temperature. At room temperature, the U-processes dominate phonon scattering. At lower temperatures, the effect of U-process become small compared to the (small) effect of defects and impurities: the scattering is reduced and the thermal conductivity is larger. At lower temperatures yet, the scattering length becomes larger than the sample size and the ability to transport heat across the sample boundaries is limited by acoustic reflection of phonons at those boundaries. The effective thermal conductivity is reduced. Thus there is a peak in the conductivity, usually well below room temperature, and the peak is sensitive to defects and impurities and to the dimensions of the crystal.

Figure 5 shows the thermal conductivity of various materials. For LiF and sapphire, the peak is nearly 200 watt/(cm-^{o}K) and the optimum temperature is 18°K and 30°K, respectively. The coolant would probably have to be liquid hydrogen. Copper would be an excellent mirror substrate material if it could be cooled (liquid helium) to about 12 ^{o}K. Natural diamond is particularly attractive because its peak conductivity of about 150 watt/(cm-^{o}K) is reached at 80°K, a temperature achievable with liquid nitrogen.

4. Synthetic Diamond as a High Power Optical Material [9]

The highest thermal conductivity can be attained if one uses synthetic instead of natural diamond. The lattice quality can be made higher and the chemical purity can in principle be made quite good. Most importantly, it can be made with a single isotope: Carbon-12. Natural carbon is composed of 98.9% carbon-12 and 1.1% carbon-13. But for optical applications, the elimination of so small a contamination has an effect on K that is far larger than its abundance would seem to indicate.

The phonon scattering by imperfections varies as the fourth power of the phonon frequency [3]. At low temperatures and for heat coupled into the material from a reservoir in contact with the surface and at a temperature near that of the diamond, the phonon frequency spectrum is thermal and the phonon scattering is low. But when energy is coupled into the bulk of the diamond crystal by the absorption of *light* (whose frequency is much higher than the phonon thermal frequency), the lattice is driven at the optical frequency and the consequent phonon scattering by the isotopic impurity is much greater. Thus, removal of the carbon-13 should result in a significant increase in the thermal conductivity of diamond at low temperature.

While this effect has not yet been measured in diamond, it has been observed in LiF and in Ge, and the magnitude of the effect in LiF is about an order of magnitude. The LiF data [4] are reproduced in figure 6a. Plotted there is the thermal conductivity as a function of temperature for concentrations of Lithium-6 that varied between .02% and 50.1%. It is apparent that the isotopically purest LiF showed thermal conductivities almost an order of magnitude larger than natural material and that the K for the purest samples may have been limited owing to the suspected presence of other impurities. The effect in diamond is expected to have a similar magnitude; figure 6b shows the measured thermal conductivity for natural diamond and the estimated values [5] for carbon-12 diamond, along with values for other materials. The K for isotopically pure diamond is expected to approach 1000 watt/(cm-$^{\circ}$K).

As indicated above, the parameter of importance is K/α . Diamond can presumably be made to have a very high K, and in a pure form it should have a low bulk optical absorption coefficient. Figure 7 compares the absorption in *natural* diamond with that of other optical materials. The absorption bands in the near infra-red are believed to be due to chemical impurities (nitrogen, mainly); the absorption at 1 micron and below should be less than 10^{-4} cm^{-1}. In a pure material, very low broad-band absorption is expected. In general, low absorption is expected for most pure single crystal materials. Figure 8 shows the absorption in sapphire [6] and in some alkali halides [6] as a function of optical wavelength. Below a few microns, the absorption is less than 10^{-4} cm^{-1}.

Thus we conclude that some single crystal materials can transmit optical powers at least 1000 times higher than glass and that diamond -- an ultraconductor of heat -- may be the best of these materials. The power-handling property will be realized only if the crystal structure is nearly perfect, the temperature is optimum, the material is chemically and isotopically pure, and if the optical absorption is low enough. Achieving these conditions, however, will require the resolution of a number of important issues.

5. Issues Associated With High Power Diamond Optics

The issues are related mainly to the use of the optic in a high power environment. Some of them are:

- Removal of heat from the optic
- Non-thermal damage
- Coatings
- Fabricability
- Material availability

In the following discussion of some of these issues, it is assumed that the application of the optic is to a powerful free electron laser device in which the electron beam is produced by an RF accelerator.

<u>Heat Removal</u> In a crystal such as diamond, energy transport is by lattice vibrations, i. e., phonons or sound waves. When the thermal conductivity is very high, the mean free path for scattering is very large -- larger than the crystal dimensions. This leads to the result that there is almost no thermal gradient in the bulk of the material. But at the edge -- that is, the interface between the crystal and the external environment -- there is a mismatch between the sound velocity in diamond and that in the external material. Phonons arriving at the interface are reflected, a process that interferes with the propagation of energy into the external world. Thus the temperature gradient is at the interface, not in the bulk material, and the difficulty associated with heat removal lies in the edge acoustic

mismatch. The process is in agreement with the acoustic model of Holland [7], which predicts that matching at the interface can be accomplished by matching the density and sound velocity of the external material to that of diamond.

There are a number of ways to ameliorate the transfer of energy through the interface. One can arrange to use a large crystal so that the large area at or near the edge compensates for the imperfect heat transfer per unit area. Alternately, one could try depositing multiple films with appropriate properties at the heat transfer interface; by gradually varying the density and acoustic velocity of each individual film, it may be possible to provide a graded interface match. Finally, heat transfer will be greatest if there is a larger temperature difference between the diamond and the external coolant; thus cooling with liquid hydrogen may be useful.

<u>Non-thermal Damage</u>   There are several sources of non-thermal damage for diamond windows: out-of-band light (i. e., harmonics of the FEL fundamental light frequency), gamma rays and neutrons produced by the stopping of the FEL electron beam, and non-linear processes such as stimulated Raman scattering induced by the relatively high peak power in the FEL micropulse.

Out-of-band Light   Free electron lasers generate some light at harmonics of the fundamental frequency. If the fundamental frequency is in the mid-infrared or near visible, light produced at the fifth harmonic and beyond will be outside the optical pass-band of the diamond and so will be absorbed near its surface. The photon energy will be above 5 eV (above the diamond band gap) and the possibility of photochemical effects and two-photon effects becomes a concern. The concern is not the heating that may be produced but the production of transient and metastable atomic and molecular species that absorb light <u>at the fundamental frequency</u> and might produce an unacceptable heat load near the surface of the optic. The same effect is likely for coatings at the surface of the optic. There is a limit to the amount of heat that even diamond with its high thermal conductivity can transport to and through its edges. For non-FEL types of lasers that operate in the visible and IR, this issue is not likely to be operative.

Neutron and Gamma-ray Damage   The electron beam of an FEL interacts with more than the undulator where the laser light is produced: a small portion of the full-energy beam may scatter and interact with the walls of the beam transport tube and the exhausted (energy-degraded) beam can interact with the material in the beam dump. Such interactions have several by-products. First, energetic photons are produced by the Bremsstrahlung process and -- if the electron energy is high enough -- by nuclear reactions that produce gamma-rays. Second, if the electron energy is above 10 - 15 MeV, nuclear reactions are possible that produce neutrons. If the fluxes of energetic photons and neutrons on the diamond optic is great enough, they can produce enough lattice dislocations to cause color centers and related effects. Color centers are likely to absorb at the fundamental frequency and produce a high heat load in the diamond. If the dislocation density is high enough, the thermal conductivity of the diamond will be affected, too.

Non-linear Processes   FELs have a pulse format that consists of a series of "micropulses" -- pulses whose duration is tens of picoseconds -- at a very high (many megaHertz) repetition rate. The peak power is much higher than the average power. For example, a laser operating at a repetition rate of 50 MHz, a pulse duration of 25 picoseconds, and an average power of 1 MW has a peak power of about 1 GW. If the beam has a diameter of 1 mm, the peak intensity I_{peak} is about 100 GW/cm^2.

At such high peak powers, the effects of a variety of non-linear processes may become important. These processes include stimulated Raman and Brillouin scattering, filamentation, whole beam self-focusing, and others. Stimulated Raman scattering (SRS) in diamond, where the SRS gain is unusually high, has been studied by McQuillan [8]. The equilibrium gain is I_{peak} * 7 cm/GW, and if the gain-length product for the duration of the pulse is large enough (say, greater than 25), a large fraction of the incident light will be shifted to a frequency about 14% lower than the incident frequency. The lost 14% appears as energy dissipated in the diamond and at high incident peak power presents an overwhelming heat load to the optic. Other non-linear processes are less important but may not be negligible.

Clearly, much work has to be done to determine whether these damage mechanisms will limit the applicability of diamond optics for high power beams. Yet there are opportunities for mitigating the effects of these processes. For example, the effect of out-of-band light can be reduced substantially by "scrubbing" the harmonics from the main beam; the effects of hard radiation can be reduced by careful design of the transport of the electron beam and by shielding of the optics; and the effect of SRS can be reduced by moving the optic a few meters farther away from the undulator of the laser.

6. Discussion

Figure 9 shows two concepts for a cooled diamond transmitting optic. The upper panel shows the use of a graded acoustic matching interface between the diamond and the cold optical mount. The coatings are selected according to the acoustic matching model and the area of the interface is made large enough to accommodate the energy flow through the interface for a small temperature difference. The coolant might be pumped liquid nitrogen (whose temperature is lower than 78 $^{\circ}$K) or perhaps liquid hydrogen. The optic is shown with a planar interface but it could just as well be a lens, grating, etc. The second panel shows a different method, one which uses a large area of interface between the diamond optic and low-pressure deposited thick films of diamond. Coupling of the energy deposited in the optic is transmitted into the diamond film readily and is presumed to be effectively coupled into the mount because of the large area of interface between the diamond film and the mount. The use of acoustic matching films would enhance the performance of this concept.

7. Summary

Cold, high-quality synthetic single-crystal mono-isotopic diamond may offer a remarkable opportunity for a very high power transmitting optic. The material is a thermal ultraconductor of heat and -- in a very pure state -- is likely to have a low absorption coefficient for light over a broad range of frequencies. Other crystal materials, such as LiF and sapphire, may offer a comparable advantage as a transmitting optic, and they are probably easier to obtain with the necessary quality. The applicability of these materials to high power lasers depends on the quantification and resolution of several issues: removal of energy through the boundaries, the actual absorption of light, damage from non-thermal and perhaps non-linear effects, and availability of material. Other issues not discussed here also require attention: suitable AR coatings, grinding and polishing, and so forth.

8. Acknowledgments

8. Acknowledgments   The author gratefully acknowledges the benefit of discussions with Russell Seitz and Jonah Jacob.

9. FOOTNOTES

* Work performed under the auspices of the U. S. Department of Energy.

10. REFERENCES

1. Gregg E. Childs et al, <u>Thermal Conductivity of Solids at Room Temperature and Below: A Review and Compilation of the Literature</u>, National Bureau of Standards, Boulder, Colorado, NBS report COM-73-50843 (1973).

2. R. Berman, <u>Thermal Conduction in Solids</u>, Clarendon Press, Oxford, (1976)

3. P. G. Klemens, "Thermal Conductivity of Pure Monoisotopic Silicon", International Journal of Thermophysics, Vol. 2, No. 4, 323 - 330 (1981).

4. Berman, Op. Cit. p. 84

5. Private Communication, Russell Seitz.

6. Science Research Corporation, <u>Novel FEL Lens</u>, final report on Contract No. DASG60-84-C-0019, Monitored by Ballistic Missile Defense Advanced Technology Center, DARPA order 4934, January, 1985

7. M. G. Holland, "Analysis of Lattice Thermal Conductivity", Phys. Rev. 132 (6) 2461 - 2471 (1963).

8. A. K. McQuillan, W. R. L. Clements, and B. P. Stoicheff, "Stimulated Raman Scattering in Diamond: Spectrum, Gain, and Angular Distribution of Intensity", Phys. Rev. A, 1 (3) 628 - 635 (1970)

9. Russell Seitz, U. S. Patent 3,895,313, July, 1975

TEMPERATURE DISTRIBUTION

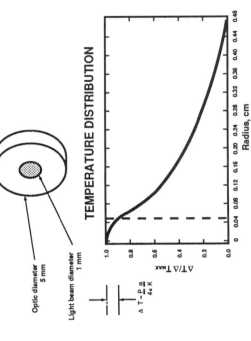

$$\Delta T = \frac{P}{4\pi} \frac{\alpha}{K}$$

Optic diameter 5 mm

Light beam diameter 1 mm

Figure 2. The temperature distribution in a slab illuminated by a laser beam. The dashed vertical line represents the edge of the illuminated area. ΔT represents the temperature difference between the center and the edge of the optical beam.

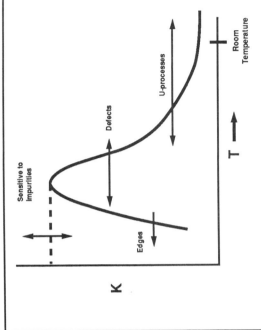

Figure 4. Dependence of K on temperature and the physical processes that are operative [2].

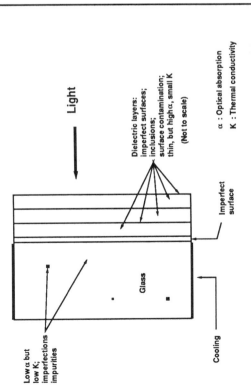

Figure 1. A typical glass AR-coated optic. Properties of the materials and their processing limit the optical power that can be transmitted through the optic.

α : Optical absorption
K : Thermal conductivity

Figure 3. The power throughput of a glass optic as a function of K/α. The range of values of K/α for a variety of suitable materials is indicated by the brackets.

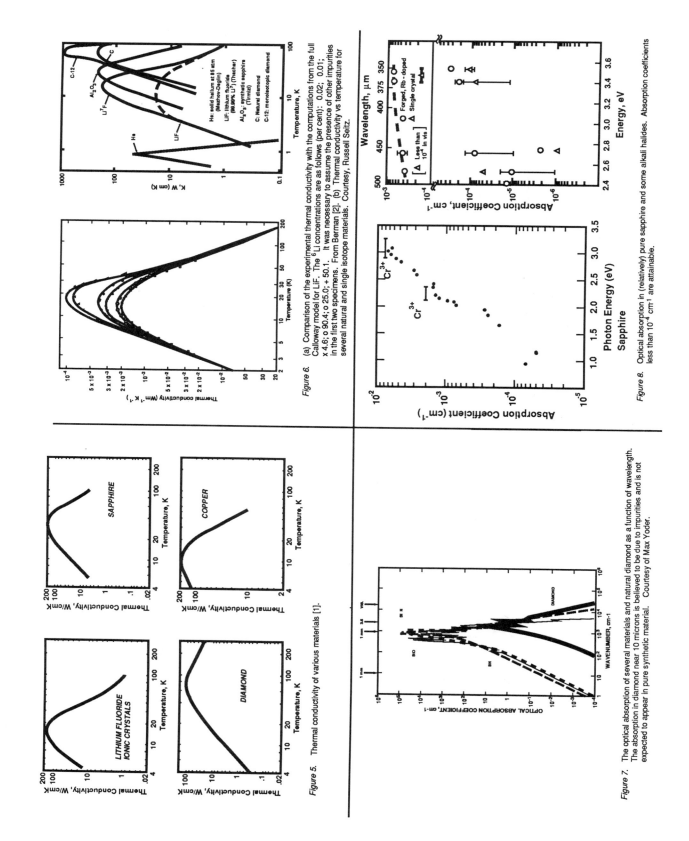

Figure 5. Thermal conductivity of various materials [1].

Figure 6. (a) Comparison of the experimental thermal conductivity with the computations from the full Calloway model for LiF. The ^{6}Li concentrations are as follows (per cent): 0.02; 0.01; x 4.6; o 90.4; o 25.0; + 50.1. It was necessary to assume the presence of other impurities in the first two specimens. From Berman [2]. (b) Thermal conductivity vs temperature for several natural and single isotope materials. Courtesy, Russell Seitz.

He: solid helium at 85 atm (Mezhov-Deglin)
LiF: lithium fluoride (99.99% ^{6}Li^7) (Thacher)
Al$_2$O$_3$: synthetic sapphire (Tirmizi)
C: natural diamond
C-12: monoisotopic diamond

Figure 7. The optical absorption of several materials and natural diamond as a function of wavelength. The absorption in diamond near 10 microns is believed to be due to impurities and is not expected to appear in pure synthetic material. Courtesy of Max Yoder.

Figure 8. Optical absorption in (relatively) pure sapphire and some alkali halides. Absorption coefficients less than 10^{-4} cm^{-1} are attainable.

Materials	n	$\frac{1}{n}\frac{dn}{dt}$	$\frac{1}{L}\frac{dL}{dt}$	$K \frac{W}{cm\,deg}$	α cm^{-1}	Topt($^\circ$K)	K/α
Quartz	1.3	-4×10^{-6}	1.3×10^{-5}	~0.01	$\sim10^{-3}\text{-}10^{-4}$	$\gtrsim$300	10-100
Quartz Crystal	1.3	-4×10^{-6}	1.3×10^{-5}	5-10	$10\text{-}10^{-4}$	12	5000-10^{-5}
Sapphire	1.6	1×10^{-5}	6×10^{-6}	100-150	10^{-4}	30	$\gtrsim 1\times10^{6}$
Li F	1.4			~150	$\lesssim 10^{-3}$	16	1.5×10^{5}
Diamond	2.42	-4×10^{-6}	$1\text{-}2\times10^{-7}$	100-150	10^{-4}	80	1.5×10^{6}
C^{12} Diamond	2.42	-4×10^{-6}	$1\text{-}2\times10^{-7}$	250-1000	10^{-4}	80	$\gtrsim 10^{7}$

Table I. A tabulation of properties for materials relevent to the transmission of high power light. T_{opt} is the temperature at which the thermal conductivity is a maximum; is the optical absorption coefficient; n is the optical index; and (1/L)dL/dT is the coefficient of expansion.

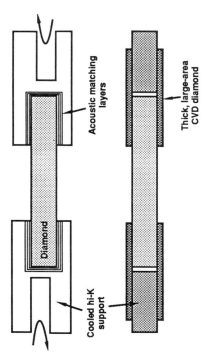

Figure 9. Two concepts for a high power diamond optic. The upper drawing shows the use of a multiple film graded acoustic matching layer at the energy transfer interface. The lower drawing indicates the use of a low pressure deposited thick diamond layer which acts as a thermal transport bridge between the diamond optic and the cooled support.

Cryogenic diamond optics for high power lasers

G. Saenz, K. Sun, R. Shah

Rocketdyne Division of Rockwell International
6633 Canoga Ave., Canoga Park, California 91303

ABSTRACT

Two concepts have been proposed for using transmissive diamond optics in the laser cavity of an FEL in order to reduce the cavity length: (1) a resonant reflector stack and (2) an intracavity divergent lens. The limitations of both designs will be discussed. Diamond was selected as the optical material for its low coefficient of thermal expansion, low thermal optic coefficient, low optical absorption coefficient, and high thermal conductivity at liquid nitrogen temperatures.

1. BACKGROUND

The free electron laser (FEL) is currently being developed as a source of high power coherent photon radiation. Despite its many attractive features, the FEL concept is necessarily plagued by the high internal photon flux and the resultant extreme thermal loads on the resonator optics. The FEL gain region requires a photon beam having small cross section and small divergence angle. Thermal distortion of the optics increases with temperature, which is proportional to the net absorbed heat flux. For a given beam energy, the heat flux is proportional to the net absorbed energy and inversely proportional to the beam area projected on the optical element. This combination leads to the need of optical elements having to withstand high fluxes of energy.

This thermal loading issue places extraordinary requirements on the resonator design, fabrication, and alignment. In fact, the FEL resonator design is primarily driven by the problem of reducing the loads caused by high photon flux. This has led to designs utilizing grazing incidence optics and long path lengths which decrease the high flux and its mirror deforming (and destructive) effects. Rocketdyne has reviewed candidate materials and concludes that diamond stands out as a promising solution to the problem of potential thermal distortion, buckling, and even breakdown, in the FEL resonator.

In making an assessment of thermally suitable optics materials, the critical factor in determining bowing and distortion is α/k, the thermal expansion coefficient divided by the thermal conductivity. Diamond's α/k is orders of magnitude lower than that of most other materials at cryogenic temperatures ($\sim 76°K$). Diamond's stiffness is also very favorable, making diamond by far the optimum substrate material for the resonator optics.

2. OPTICS DESIGNS

Two concepts have been proposed for using transmissive diamond optics in the laser cavity of an FEL in order to reduce the cavity length. These are the resonant reflector stack and the intracavity divergent lens. Diamond has been selected as the optical material because it is tough and, at liquid nitrogen temperatures, has an extremely high thermal conductivity coupled with a very low coefficient of thermal expansion and a low thermal optic coefficient. There is also some indication[1] that the intrinsic absorption in diamond at 500 nm may be as low as 3×10^{-4} cm^{-1} .

The first proposal, the resonant reflector stack, can be viewed as an attempt to obtain the anticipated benefits of a diamond multilayer coating from a macrolayer stack of thin diamond shells. Although the diamond shells would be thick enough to be freestanding, the reflections from the uncoated surfaces would be made to constructively interfere by accurately holding the diamond shell optical thickness to (N+1/4) times the nominal (center) wavelength of the laser light, where N is a large number. This resonant enhancement of the reflections from the stack would have to be maintained under thermal operating stresses by controlling the shell thicknesses as well as the separation between the shells to a small fraction of a wavelength. The diamond shells would be edge cooled or, for higher energy density requirements, be sufficiently separated for a liquid or gaseous coolant to flow between the shells (face cooled) to carry away the heat energy resulting from absorption. The high index of refraction (2.43 at 500 nm) allows one to obtain a high reflectivity using just a few shells.

The alternative proposal is to place a diamond lens of negative power in the FEL cavity. This lens would increase the beam divergence significantly above that due to diffraction so that the flux density would decrease more rapidly along the optical axis. In this way, a reflecting mirror with a given damage threshold could be moved much closer to the wiggler. The lens could be edge cooled or, with a little more complexity, face cooled to remove the heat from absorption.

We have examined both proposals, attempting to identify those factors affecting the viability of each concept. Extremely short FEL pulses are found to impose significant limitations on the allowable resonant reflector stack designs.

2.1. Resonant reflector stack

We have investigated the reflectance of a stack of diamond shells separated by vacuum. Virtually the same results would be found if a gas replaced the vacuum, since the refractive indices of transparent gasses under normal pressures are close to one. Slightly lower reflectance would be found if liquid nitrogen (n = 1.22) were separating the shells.

Because of diamond's high index, reflectance can exceed 99.9 percent with just five shells. For just two shells, the reflectance is 0.892, if the quarter wave condition is satisfied in both the shells and their separations. The corresponding values for 3, 4, 5 shells are 0.981, 0.9967, and 0.9994, respectively.

One might imagine a typical five-shell stack as consisting of five diamond shells each 0.5 mm thick, separated by 1 mm spaces. This configuration is illustrated in figure 1.

This arrangement will <u>not</u> function as a resonant stack for very short duration laser pulses. For example, a 20 picosecond pulse has a length of only 6 mm. The optical path through this stack and back is equivalent to over 20 mm in vacuum. By the time the leading edge has returned to the front of the stack, the trailing edge of the pulse reflected from the front of the stack is already gone so that interference between the two is not possible.

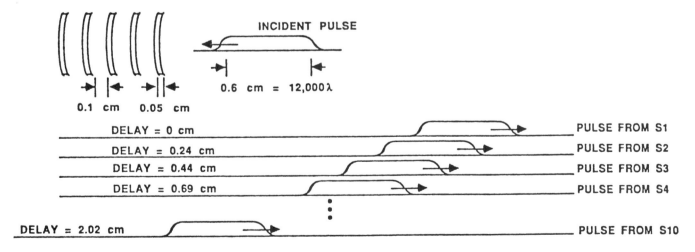

THICK STACK: 5 SHELLS 0.05 cm THICK, WITH 0.1 cm SPACING BETWEEN SHELLS

- **LITTLE OR NO INTERFERENCE BETWEEN REFLECTED PULSES FROM NON-ADJACENT SURFACES**
- **TOTAL REFLECTED PULSE STRETCHED FROM 0.6 cm TO MORE THAN 2 cm.**

Figure 1. The relative positions of the reflected pulses from a thick resonant reflector indicate the spatial overlap is minor. Hence, the high reflectivity possible with a five-shell stack cannot be realized with such a short incident pulse.

We can insure that most of the pulse experiences self-interference by making the stack thin compared to the pulse length. We expect that a stack which is 5 percent of the pulse length (0.3 mm) will behave like a coherent reflector. This restriction introduces two possible problems. The diamond shells may have to be so thin that they are no longer robust, and they may have to be so closely spaced that they would have to rely on edge cooling since the flow of coolant through them would be severely restricted. These restrictions may render this approach nonviable for short pulse systems.

The bandwidths of resonant reflectors are extremely narrow[2] and, therefore, the spectral width of the incident light must also be considered. Because the spectrum is the Fourier transform of the wavetrain, no pulse of light can be rigorously monochromatic. For the 20 ps pulse (6 mm, or 12,000 wavelengths) considered above, the spectral width $\Delta\lambda/\lambda$ must be of the order of 10^{-4}, and it may be greater if the pulse is chirped or spectrally broadened. Therefore, the quarter wave resonance condition cannot hold exactly for all the frequencies in the pulse. For two wavelengths differing by 0.01 percent, requiring that the phase change of each upon traversing the stack be no greater than $\lambda/10$ imposes a limit of $10^3\lambda$ on the optical path length in the stack. This restriction is of the same order as the first one, namely, a reflector stack which is thin enough to ensure self-interference will also be resonant for all wavelengths in the pulse, unless the frequency spread is much worse than what the uncertainty principle requires.

We have used FILMSTAR (a commercially available thin film analysis program) to calculate the reflectance of a number of diamond reflector stack designs employing three, four, or five diamond shells. Typical results are presented in figure 2. The shells in all three stacks have an optical thickness of 160.25 wavelengths (which corresponds to a physical thickness of ~33 μm), and the shells are separated by 64.25 wavelengths (~32 μm).

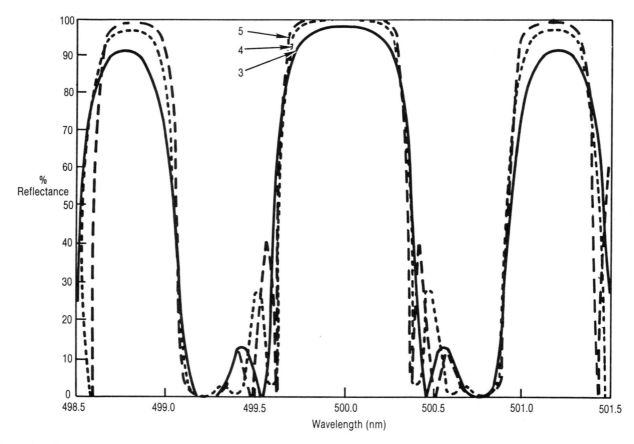

Figure 2. The spectral reflectance for resonant reflectors composed of three, four, and five diamond shells is shown. Note the widths of the high reflectance bands, which are quite narrow despite the fact that the shells are only ~33 μm thick, quite thin for a resonant reflector.

These stacks are thin enough to satisfy the conditions we derived above; nonetheless, the reflectance peak is still quite narrow. Just 0.01 percent away from the nominal wavelength of 500 nm the reflectance is at most a few tenths of a percent less than its peak value. At wavelengths 0.1 percent from nominal, however, the reflectance has fallen precipitously, to under 20 percent.

This suggests that a longer pulse length aids in the viability of this design in two ways. It allows for the possibility of greater monochromaticity (by the uncertainty principle) and this allows for thicker stacks with greater separations (which results in a narrower resonance). It also allows for the longer optical path length in the stack to satisfy the interference requirements.

The close spacing of the shells indicates that heat removal must be accomplished predominantly through edge cooling rather than face cooling. The expected low value of dn/dT (it is ~10^{-5}/K at room temperature), in conjunction with the thinness of the shells required by the considerations described above, indicates that a significant temperature rise may be accommodated. FILMSTAR results indicate changes in the shell optical thickness of less than 0.05 wavelengths are acceptable, particularly if the spacing between shells can be adjusted to partially compensate for the optical path change in the shell. A thermal analysis is required to determine if the temperature distribution resulting from absorption and edge cooling is sufficiently uniform and does not cause the $\lambda/20$ limit to be exceeded.

2.2. Intracavity negative power lens

The second concept involves inserting a diamond lens of negative power into the FEL cavity to increase beam divergence. A more divergent beam permits a cavity mirror with a given damage threshold to be placed closer to the wiggler, shortening the cavity length. Obviously, the lens is the critical element in this scheme, as it must transmit a much higher power density than that reflected by the mirror behind it. This implies a suitable combination of low absorption and high thermal conductivity, as well as a means of lowering the reflection loss at the diamond/vacuum interface.

Because diamond has a high index of refraction ($n_D = 2.43$), the reflectance of an uncoated diamond surface is large, about 17.4 percent at normal incidence. Thus, in traversing the two uncoated surfaces of a simple lens, about 30 percent of the laser power would be lost to Fresnel reflection. Normally, one lowers the reflectance value by coating the surface. However, coatings generally have significantly lower damage thresholds than the same material in bulk form. Indeed, one of the implied assumptions prompting us to investigate diamond optics concepts was that optical coatings could not withstand the extremely high average and peak power densities in the FEL cavity we had in mind. If such coatings ever are developed, it would be much more logical to employ them in a high reflectance coating on a diamond substrate and thereby avoid the thermal problems inherent in transmissive optics.

If the diverging lens could be inserted at Brewster's angle in the cavity, the coating requirement could thereby be avoided. The geometry of this arrangement is illustrated in figure 3. Brewster's angle at the vacuum-diamond interface is about 67.6°, so the lens would have to be specifically designed for this large tilt. Since at least one surface must be curved, the Brewster's angle condition cannot be satisfied everywhere on both surfaces of the lens. A complicating factor is that Brewster's angle at the diamond-vacuum interface, 22.4°, is close to the critical angle of 24.3°, at which total internal reflection occurs. To avoid the high reflectance losses near the critical angle, the curvature of the lens surface, and hence the lens power, must be limited. Using the standard formula, we find that the reflectance remains below 0.3 percent for angles of incidence from 0.5 degrees below the Brewster's angle to 0.4 degrees above Brewster's angle. Therefore, if we restrict the curvature of the second surface so that the angle any light ray makes with the surface normal at its point of intersection will be within $\pm$ 0.4 degrees of Brewster's angle, the total reflection loss will be much less than one percent. We can say much less because while the outer rays will experience reflections of about 0.3 percent, near the center of the beam where the power density is much greater, the reflectance is much less.

The beam incident on the curved diamond-vacuum interface will be polarized exactly parallel to the plane of incidence only along that line which is perpendicular to the tilt axis and runs through the center of the surface. Elsewhere the incident beam will possess a small component of perpendicular polarization, with E_σ on the order of $0.007E_\pi$. This polarization is strongly reflected near Brewster's angle ($R_\sigma \approx 51$ percent). However, since only about 5×10^{-5} of the energy is in the orthogonally polarized component, this reflection loss can be neglected.

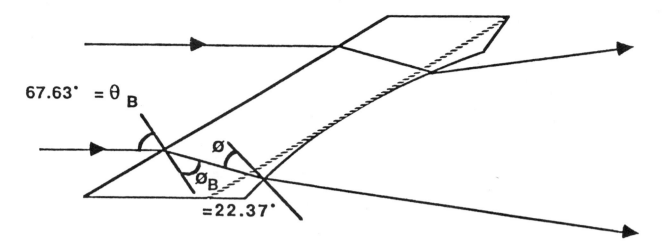

$67.63° = \theta_B$

ϕ

ϕ_B

$=22.37°$

Figure 3. The geometry of a plano-concave lens inserted at Brewster's angle is indicated. The curvature of the concave surface has been exaggerated for clarity; significant departures from Brewster's angle on the concave surface carry the penalty of large reflection losses.

The results have been summarized in the table below. These results are prescriptions for three lens designs.

Laser beam radius (cm)	Clear aperture radius (cm)	Radius of curvature (cm)	Focal length (cm)	Sag (μm)
0.1	0.26	37	-2.9	9
0.2	0.53	76	-5.8	19
0.4	1.05	150	-11.6	37

Table 1. Parameters for a lens at Brewster's angle, based on an angle of incidence for the edge ray that is 0.4° from Brewster's angle.

The relatively short focal lengths obtained cause a rapid increase in the cross section of the refracted beam. The beam diameter will increase ten times in a distance of nine focal lengths, so the incident beam power density will have decreased by a factor of 100 in this distance.

This performance is impressive, but one should keep in mind that these results come from a first order analysis. Because of the large incident angles (~22°), it is possible that significant beam aberrations will be present. Very likely these can be corrected by generating an aspheric surface rather than a sphere on the lens, but any definitive pronouncement must await a detailed analysis of this rather unusual lens.

3. THERMAL ANALYSIS

Our thermal analysis considered a diamond disk inserted in the cavity normal to the laser beam. Hence, we cannot apply those results directly to the Brewster's angle lens. Some features of the Brewster's angle lens would tend to make it perform more poorly in a thermal sense. The greater angle of incidence gives an eight percent longer ray path in the lens for a given thickness, and the large inclination to the beam axis results in a conduction path from the lens center to its edge that varies from 100 percent to 260 percent of the length of the conduction path in a normal incidence lens. On the other hand, the large angle of incidence also spreads the incident power over a larger surface area, reducing the flux density to just 40 percent of its value for a normal incidence lens. Also, the obvious possibility of face cooling for this design would increase the power handling capabilities significantly.

Since the most stringent power dissipation concept relies on edge cooling the lens (as opposed to face cooling) to remove the absorbed energy, the following discussion will describe this "lower limit" of power handling capability for both designs. The steady state temperature distribution in a normal incidence diamond window has been calculated using a temperature-dependent thermal conductivity.

We have analyzed the situation of a diamond disk of uniform thickness which, except for an outer annulus, is illuminated by a normally incident, collimated, truncated Gaussian laser beam centered on the disk. The disk is assumed to be optically thin, and the absorption coefficient is assumed fixed and characteristic of the bulk material. There is no surface contribution to the absorption. Reflections at either surface are ignored. Together, these assumptions lead to a distribution of heat sources within the window that is independent of the coordinate normal to the disk surface.

All cooling is assumed to take place via conduction to the edge of the disk, which is held at a fixed temperature. The value of the thermal conductivity at any point in the diamond window will not depend explicitly on the coordinates of the point but will depend on the local temperature according to the equation

$$k(T) = (3 \times 10^4) \, T^{-1.25} \quad (\text{W/cm-K}). \tag{1}$$

This expression matches the best measured values for type II diamond fairly well from 80 K to 500 K. At higher temperatures, the conductivity of our model does not drop quite as rapidly as the measured values, so that at 1800 K the model value is about 50 percent too high. This is the highest temperature of interest to us, since at 1800 K graphitization proceeds relatively rapidly. As a consequence of our assumptions, the heat flow in the window is purely radial, and the temperature distribution is identical in every plane parallel to the window faces.

The heat generated at any point in the diamond window, h(r), can be straightforwardly calculated from the power density in the truncated Gaussian beam. Knowing h(r) and k(T), one can then solve the heat flow equation,

$$h(r) = -k(T) \, (dT/dr), \tag{2}$$

for the steady state temperature distribution T(r). We did not have to solve this equation numerically; the solution was found analytically in terms of the exponential integral function. The details of this calculation are presented in a separate report on this thermal model.

4. THERMAL MODEL RESULTS

The model described in the previous section was used to predict the steady state temperature distribution in a diamond window with a radius of 0.6 cm illuminated by a Gaussian beam truncated at a radius of 0.49 cm and with a 1/e radius of 0.2 cm. The average power in the beam was chosen to be 20 MW, and results, plotted in Figure 4, were calculated for five values of the absorption coefficient ranging from 3×10^{-2} cm^{-1} down to 3×10^{-4} cm^{-1}. The absorption coefficient and the incident power enter the model only as a product. Therefore, the temperature distribution shown for $\alpha = 3 \times 10^{-4}$ cm^{-1} and a power of 20 MW is identical, for example, to what would be found for an absorption of 3×10^{-2} cm^{-1} and a power of 0.2 MW.

The results are independent of thickness because of our assumptions that only bulk absorption is present, the window is optically thin, and all the heat flows out through the cooled rim of the window. But while the temperature distribution is unchanged, the power which is absorbed in the window increases in proportion to its thickness. We have assumed that the window edge has a fixed temperature of 80 K. The load on the cooling system to maintain this temperature increases in proportion to the window thickness.

We see from Figure 4 that the window undergoes thermal runaway at $\alpha = 10^{-2}$ and 3×10^{-2} cm^{-1} (average power = 20 MW). An absorption of 3×10^{-3} cm^{-1} results in a temperature rise on the beam axis of 320 K. Absorption values of 10^{-3} and 3×10^{-4} cm^{-1} produce more reasonable maximum temperature rises in the window of about 50 K and 12 K respectively. We note that if the window is 0.1 cm thick and is constructed of diamond with an absorption coefficient of 10^{-3} cm^{-1}, then the heat load on the cooling system is 2 kW.

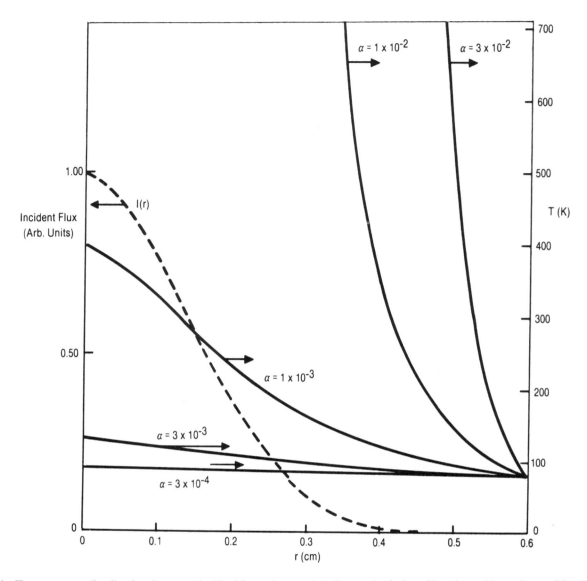

Figure 4. Temperature distribution in an optically thin, edge-cooled diamond window illuminated by a P_{av} = 20 MW truncated gaussian laser beam. These results were derived using a temperature-dependent thermal conductivity.

The appropriate value of the thermal conductivity to use in calculating the window thermal performance may not be as high as the values we have assumed[3], since these high values at cryogenic temperatures are affected by the geometry of the conducting sample[4] as well as any departures of the phonon momentum and energy distributions from thermal equilibrium. On the other hand, it is likely that eventually synthetic diamonds of higher quality (fewer defects and impurities) than the best natural diamond will be produced, and these should show higher conductivities attributable to the reduced number of scattering centers for phonons. In addition, in the production of synthetic diamond the opportunity exists for producing a crystal from a single isotope resulting in an additional reduction in the number of scattering centers. The question of whether these high conductivity values can be realized in the thin diamond transmissive optical elements under consideration here is a complicated one and awaits experimental measurement.

Finally, we note that experimentally it has been observed that phonon scattering from the diamond surfaces has a specular component[5], and conductivities greater than what would be obtained with the completely diffuse reflection of phonons have been measured. Since the diamond optical elements in a laser cavity would be very highly polished, they may exhibit a higher conductivity than one would expect from simple size considerations alone.

5. SUMMARY

The short pulse length is the critical factor affecting the design of the resonant reflector, since it limits the thickness of the stack reflector and thus prevents any significant coolant flow between the shells for face cooling. If the FEL pulse could be lengthened, both the fabrication and cooling problems associated with the reflector stack would be eased. Thicker shells would also be less fragile. However, if the shells in the stack were made M times as thick, the already narrow high reflectivity band would shrink to roughly $1/M$ times its original width. (The reflectance of the design of Figure 2 falls to 50 percent for a fractional change of $\sim 7 \times 10^{-4}$ in the wavelength.) If the spectrum of the pulse did not narrow correspondingly as the pulse lengthened, then the thicker stack would be unable to offer high reflectance to all the frequency components in the pulse. The reflectance decline with increasing angle of incidence is also more rapid as the stack thickens, making alignment more critical. Finally, the optical thickness of thicker shells is more sensitive to a temperature change. Any design change to permit the diamond shells to be made M times as thick will accommodate a temperature rise of only $\Delta T/M$, where ΔT is the temperature rise accommodated by the reference design. Therefore the cooling requirements increase faster than, not proportional to, the increase in thickness.

An intracavity lens when employed at normal incidences suffers from unacceptably high insertion loss due to Fresnel reflection from its two surfaces. However, we found that a lens at Brewster's angle could reduce reflection loss below one percent and still possess sufficient power to increase the divergence of the beam a practical amount. The aberrations introduced by the large tilt have not been examined; their magnitude and the difficulty expected in correcting or compensating them are the remaining major optical issues that must be examined.

Both designs will function at higher power than the lower limit analysis defined by edge cooling if face cooling can be achieved; however, the difficulties in propagating a high photon flux through liquid nitrogen have yet to be examined. The viability of these two concepts for diamond optics also rests in large part on the possibility of being able to produce low absorption synthetic diamond.

6. REFERENCES

[1] G.A. Slack, General Electric Physical Chemistry Laboratory, Report No. 79CRDO71, June 1979.

[2] J.A. Dobrowski, "Coatings and Filters" in Handbook of Optics, W.G. Driscoll, ed. (1978), p 8-67.

[3] R. Berman, "Thermal Properties" in The Properties of Diamond, J.E. Field, ed. (1979), p 12.

[4] N.W. Ashcroft and N.P. Mermin, Chapter 25 in Solid State Physics (1976).

[5] R. Berman, "Thermal Properties" in The Properties of Diamond, J.E. Field, ed. (1979), p 15.

LASER DAMAGE OF DIAMOND FILM WINDOWS

S. Albin, A. Cropper and L. Watkins
Department of Electrical and Computer Engineering
Old Dominion University, Norfolk, VA 23529-0246

C. E. Byvik and A. M. Buoncristiani
NASA Langley Research Center, Hampton, VA 23665-5225

K. V. Ravi and S. Yokota
Crystallume, 125 Constitution Drive, Menlo Park, CA 94025

The many unique physical properties of diamond make it useful as a thin film coating for laser optics. We have calculated the laser induced thermal stress resistance for diamond and other optical materials. The calculated stress resistance for diamond is orders of magnitude higher than any other material and, therefore, diamond films should have a higher laser damage threshold. Calculations also indicate that diamond film, because of its high thermal conductivity, exhibits tolerance for isolated impurity inclusions. Polycrystalline diamond films were deposited on silicon substrates using a d.c. plasma enhanced chemical vapor deposition process. The films were characterized by Raman and optical absorption spectroscopy and by ellipsometry. Laser induced damage thresholds of diamond film windows and films on silicon substrates were measured for single pulses of 532 nm and 1064 nm laser radiation. The measured damage thresholds for diamond windows are 6.0 J/sq.cm (300 MW/sq.cm) at 532 nm and 12.4 J/sq.cm (620 MW/sq.cm) at 1064 nm. For diamond on silicon, the damage thresholds are 3.65 J/sq.cm (182 MW/sq.cm) at 532 nm and 14.4 J/sq.cm (720 MW/sq.cm) at 1064 nm. These values compare favourably with those for other common materials used as optical coatings.

1. INTRODUCTION

Diamond has many unique properties distinguishing it from other solid state materials. Its thermal, mechanical and optical properties are superior to those of other widely used semiconductors. Many important properties of diamond are shown in Table 1 and compared with those of other group IV elemental semiconductors and gallium arsenide. As is evident, the thermal conductivity, hardness and breakdown field are several times higher than the other materials listed, suggesting resistance to damage from intense optical radiation. These properties of diamond have already been effectively exploited to develop electrical and optical devices.[1-3] The wide bandgap and stable color centers in diamond have also been used to make tunable solid state lasers.[4] However, the lack of large scale economical production of high quality films has limited wide spread application of diamond films.

Chemical vapor deposition of diamond films using reactive pulse plasma, hot filament, electron bombardment and d.c. plasma have been reported.[5-11] With the progress achieved in chemical vapor deposition of diamond films,

Table 1
Comparison of important properties of diamond
with other semiconductors

Properties	Ge	Si	GaAs	Diamond
Band Gap (eV)	0.66	1.12	1.43	5.45
Breakdown Field (V/cm)	$\sim 10^5$	5×10^6	6×10^6	$>10^7$
Carrier Lifetime (s)	2×10^{-4}	2.5×10^{-3}	10^{-8}	$\sim 10^{-10}$
Dielectric Constant	16	11.8	13.1	5.5
Electron Mobility (cm^2/V-s)	3900	1500	8500	1900
Electron Velocity (cm/sec)	6×10^6	1×10^7	2×10^7	2.7×10^7
Hardness (Kg/mm^2)	780	10^3	600	10^4
Hole Mobility (cm^2/V-sec)	1900	600	400	1600
Lattice Constant (A^0)	5.64	5.43	5.65	3.57
Melting Point (^{0}C)	941	1420	1238	~ 3800
Refractive Index	5.6	3.4	3.6	2.4
Resistivity (Ohm-cm)	43	2.5×10^5	4×10^8	$>10^{16}$
Thermal Cond.(W/cm-^{0}K)	0.64	1.45	0.46	20
Thermal Expansion Coeff.	5.5×10^{-6}	2.6×10^{-6}	5.9×10^{-6}	8×10^{-5}

new applications of diamond films in laser optics are possible. Since surfaces of optical elements are often most sensitive to laser damage, increasing the surface damage threshold with an optical coating can be expected to improve the damage threshold of these elements. High reflection and anti-reflection coatings on optical elements in laser systems are used to optimize performance. These films most often are the weak elements which limit the energy flux from a laser. Improvements in the laser damage threshold, E_d, of these films will significantly reduce design requirements as well as increase the transmitted laser power limits. Conversion efficiency of nonlinear crystals such as KDP or KTP increases with input power but the low damage threshold of these crystals prevent their optimum use. If damage threshold can be increased by coating the crystals with appropriate films then higher conversion efficiencies can be obtained.

2. THERMAL STRESS RESISTANCE PARAMETER

One measure of the laser damage tolerance of a material is the laser induced thermal stress resistance parameter, R_T, defined by the following expression,

$$R_T = \sigma_f K(1-\nu)/\alpha E \qquad (1)$$

where σ_f is the tensile fracture strength, K is the thermal conductivity, ν is Poisson's ratio, α is the thermal expansion coefficient and E is the elastic modulus. R_T has been used as a figure of merit for evaluating laser materials tolerance to laser damage.[12] The higher the thermal stress resistance parameter of a material, the higher its laser damage threshold. We have calculated R_T for diamond as well as other common laser materials; the results are shown in Fig. 1 as a plot of log K versus log R_T.

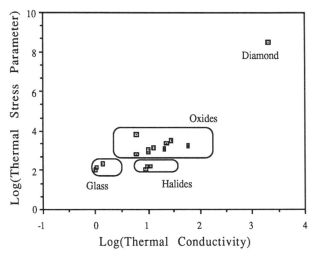

Figure 1. Thermal Condutivity, K versus Thermal Stress Parameter, R_T.

The thermal conductivity and thermal stress resistance of similar materials cluster together in groups. The oxide group includes sapphire, spinel

and commonly used garnets (YAG, GSGG). The halide group includes fluorides and chlorides of magnesium, calcium and lithium and the glass group includes phosphate and silicate glass. Diamond, having a thermal conductivity several times higher and a thermal stress parameter orders of magnitude higher than the other laser materials, appears to be a good choice as a material tolerant to laser damage. Diamond films may be used for a variety of optical coatings as well as for optical windows. Therefore, it is important to measure the laser damage threshold of diamond films.

3. LASER INDUCED DAMAGE
3.1 Dielectric Breakdown

If the damage is due to dielectric breakdown induced by the laser radiation, the laser power density, P_d, is related to the dielectric breakdown field, V_b, by the equation,

$$P_d = V_b{}^2 \, n/Z_o \qquad (2)$$

where n is the refractive index and Z_o is the impedance of free space. For bulk diamond, Equation (2) gives $P_d = 600$ GW/cm². For 10.6 μm laser radiation a threshold of 4 GW/cm² has been measured for bulk diamond.[13] In the case of thin films of high bandgap materials, the Forlani-Minnaja law[14] predicts $V_b \propto d^{-1/2}$ where d is the film thickness. Therefore, diamond thin films should have a high breakdown field and a high laser damage threshold.

3.2 Damage due to Impurities

When the films contain impurities and inclusions which absorb the laser radiation the damage is caused by the thermal stress induced by laser heating. The temperature rise of the host due to a spherical impurity absorbing radiation as shown in Fig. 2 is given by[15,16]

$$T_p = \frac{a^2 QI}{K_p} \left[\frac{1}{3} \frac{K_p}{K_h} + \frac{1}{6} \left(1 - \frac{r^2}{a^2} \right) - \right.$$

$$\left. \frac{2ab}{r\pi} \int_0^\infty \frac{e^{\frac{-y^2 t_p}{\gamma_1}}}{y^2} \frac{(\sin y - y \cos y)(\sin(ry/a)) \, dy}{(c \sin y - y \cos y) + b^2 y^2 \sin^2 y} \right]$$

K_p and K_h are the thermal conductivity of the impurity and host respectively, a is the radius of the impurity, Q is the absorption cross section, I is the laser intensity, t_p is the laser pulse duration and $b = [(K_p^2 D_h)/(K_h^2 D_p)]^{1/2}$,

$c = 1-(K_h/K_p)$ and $\gamma_1 = a^2/D_p$, where D_p and D_h are the thermal diffusivity of the impurity and the host. The above equation can be solved for the laser energy required for the melting points of the impurity or the host.

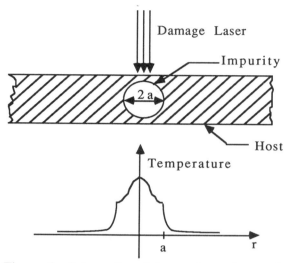

Figure 2. Schematic of laser absorption and temperature rise in a host containing an isolated impurity

In Fig. 3 we show the variation of threshold energy density, required for breakdown at a = r, with K_h when various isolated impurities are present in the host. It is clear that the damage threshold of diamond is not severely affected by the impurities. Also for r>>a, the calculated value of E_d for diamond film was found to be a constant, independent of the thermal conductivity of the impurities.

4. EXPERIMENT

The diamond films were prepared by a plasma enhanced chemical vapor deposition (PECVD) process. Typical range of conditions employed were as follows:

Temperature:	600 to 800°C
Pressure:	20 to 30 Torr
CH_4/H_2:	0.1 to 5

Silicon wafers were used as substrates. Free-standing diamond film windows were produced by etching back the substrate. Laser damage on the samples was induced by varying the energy from 1 to 100 mJ from a 1064 nm Nd:YAG laser with a pulse duration of 20 ns. Silicon, diamond film on silicon and diamond film

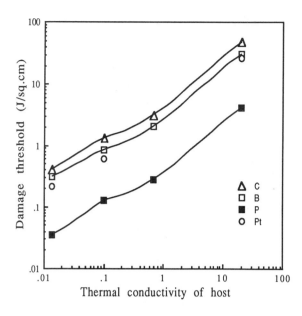

Figure 3. Variation of damage threshold with thermal conductivity of hosts containing impurities.

windows were used as samples. The diameter of the laser spot at the impact point was measured using an array detector. The laser damage threshold was measured using the setup shown schematically in Fig. 4. The sample was mounted on an X-Y-Z microposition stage. A He-Ne laser was used as a probe to measure the reflectance of the surface.

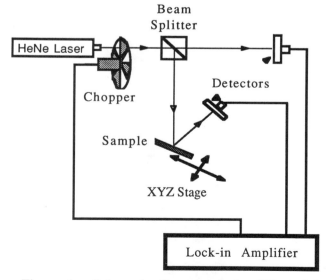

Figure 4. Schematic diagram of the differential reflectometer used in the damage threshold measurements.

The reflection from an undamaged portion of the surface was detected and fed to the lock-in amplifier along with the reference signal to obtain a null point. The probe was then positioned on the damaged spot and the intensity of the reflected beam was measured. A deviation from the null condition occurred when the reflected intensity changed, due to absorption and scattering from the damaged spot. The deviation was measured as a function of laser damage energy. The damage on the films was also confirmed by using a low power optical microscope. The experimental set-up could be easily modified to detect the scattered light by blocking the specular reflection. The refractive index of the film was measured using an ellipsometer and the films were analyzed using Raman and absorption spectroscopy and scanning electron microscopy.

5. RESULTS AND DISCUSSION

All the diamond film samples studied were polycrystalline. The front surface was faceted whereas the back surface (etched-back) was smoother than the front as shown in the SEM photograph (Fig. 5a & b). The corresponding Raman spectra of these surfaces are shown in Fig. 5c & d , however, there is no significant differences between the spectra from the front and back surfaces of the film. All the diamond films showed characteristic first order Raman spectral line around 1332 cm^{-1}, which may be compared with the spectrum shown in Fig. 5e for a natural bulk diamond. The significant

(a)

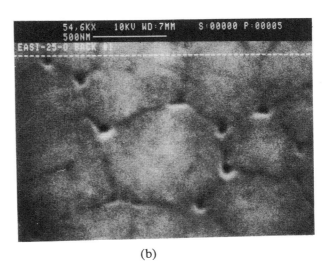

(b)

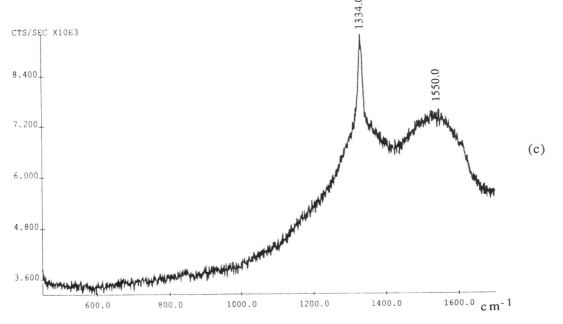

(c)

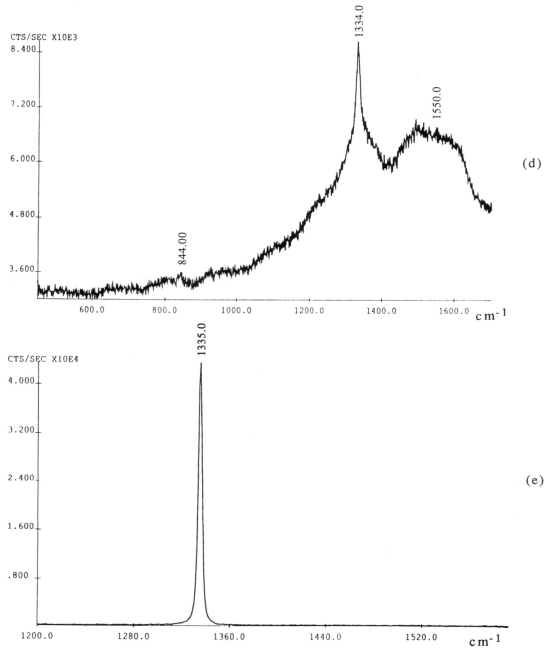

Figure 5. (a) and (b) SEM of the front and back of a diamond film window,
(c) and (d) corresponding Raman spectra and (e) a Raman
spectrum of a bulk diamond sample.

difference between the spectra is a broad fluorescence background centered around 1550 cm^{-1} for the diamond film, which is an indication of the presence of sp^2 bonding due to graphitic carbon in diamond film. The refractive index of various samples, measured using ellipsometry, was in the range of 2.6 to 2.7 which is higher than the bulk value of 2.41. This is probably due to graphitic inclusions in the film and also due to surface roughness.

The absorbance spectrum of diamond films was measured using a Perkin Elmer spectrophotometer and is displayed in Fig. 6, for the wavelength range from 500 to 1500 nm. Interference effects due to the film (1.87 μm) can be clearly seen. No correction for Fresnel reflectance from the film has been made in the absorbance spectrum shown. Absorption of the film increases towards shorter wavelengths. The absorption coefficients derived from Fig. 6 by applying the corrections for reflectance are α_{532} = 3.93 x 10^3 cm^{-1} and α_{1064} = 1.20 x 10^3 cm^{-1}.

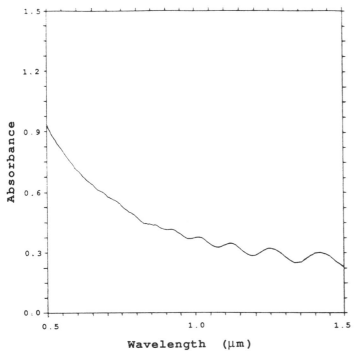

Figure 6. Absorbance spectrum of 1.87 μm diamond film.

In Fig. 7a and b we show the results of laser damage on silicon substrate, diamond film on silicon and diamond film windows for 532 and 1064 nm laser radiation respectively. The output signal from the lock-in amplifier is plotted against the laser energy. The damage threshold of silicon was measured to be 5.3 J/cm^2 (265 MW/cm^2) for 1064 nm and 2.1 J/cm^2 (105 MW/cm^2) for 532 nm laser pulses. These are within the range of values reported for silicon.[13] The irradiated spots were also examined using a low power optical microscope for laser indced damage and the threshold energy agreed with that determined by the reflectance technique. The absorption coefficients of silicon at these wavelengths are high and energy transfer by resonant surface plasmons has been considered as a damage mechanism.[17] The measured damage thresholds for a silicon substrate coated with a 1.87 μm diamond film are 3.65 J/cm^2 (182 MW/cm^2) at 532 nm and14.4 J/cm^2 (720 MW/cm^2) at 1064 nm. The reflectance of silicon substrate is about 30% and the 1.87 μm diamond film corresponds to approximately an equivalent optical thickness of quarter wavelength at 532 nm. For this

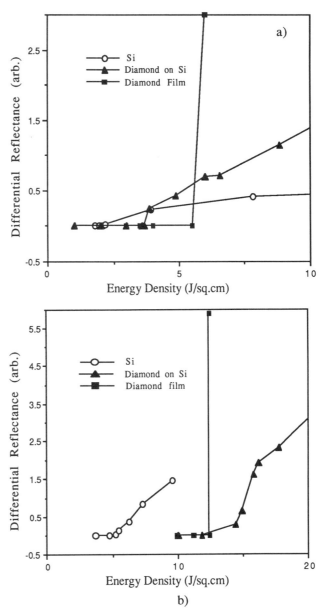

Figure 7. Differential reflectance as a function of laser energy density of (a) 532 nm and (b) 1064 nm.

film-substrate combination, the reflectance is reduced to approximately 8%. Assuming the laser damage threshold of diamond to be higher than silicon, it is reasonable to expect a threshold of 1.6 J/cm^2 for this film-substrate combination, if the damage occurs mainly at the substrate. The measured damage threshold at 532 nm is higher than that of the substrate, showing the effectiveness of a diamond film for laser hardening. In the case of 1064 nm laser radiation, the same diamond film has an optical thickness of non-quarter wavelength; hence the reflectance of the film-substrate combination is between 8 and 30%. Morever, for a non-quarter wavelength film the maximum electric field due to the laser radiation will be within the film rather than at the interface. Such a design has been shown to be beneficial in achieving a high damage threshold for optical coatings.[18] The high damage threshold of 14.4 J/cm^2 at 1064 nm obtained for the film-substrate combination is an indication that diamond film will be useful for a variety of optical coating applications involvng high power lasers.

For a diamond film thickness of 1.87μm, the laser damage thresholds at 532 and 1064 nm were found to be 6.0 J/cm^2 (300 MW/cm^2) and 12.4 J/cm^2 (620 MW/cm^2) respectively. These values are substantially higher than those measured for the silicon substrate. Thus, the low damage threshold of the film-substrate combination discussed above for 532 nm is not due to the diamond film. However, the damage threshold measured for free-standing diamond film is lower than theoretically predicted value. Since the films have high absorption coefficient due to sp^2 bonded carbon the contribution from these impurities will be substantial in determining the damage threshold. However, the damage threshold may be further improved by optimizing deposition conditions.

In Fig. 8 we show the SEM photographs of laser damage. There is an obvious difference in the nature of damage on silicon substrate and on film-substrate combination. Surface melting is visible on silicon whereas the damage on the latter appears to be due to dielectric breakdown. The diamond films developed cracks during laser damage suggesting that the film stress may influence the damage threshold.

6. SUMMARY

Free-standing diamond films have a high laser damage threshold at 532 and 1064 nm; therefore they are useful as laser windows and as an overcoat for high reflectors and undercoat for antireflectors. In combination with suitable substrates, diamond films can be an excellent antireflection coating for infrared windows. However, the experimentally measured damage threshold was found to be lower than the calculated values. Dielectric breakdown induced by the laser radiation seems to be responsible for the damage in diamond films, though the thermal stress induced by highly absorbing sp^2 bonded carbon impurities may be a contributory factor. Deposition conditions as well as stress of

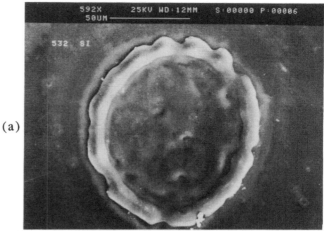

(a)

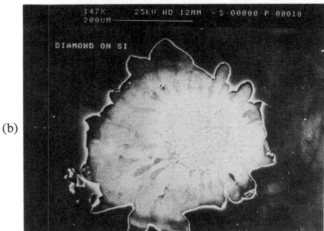

(b)

(c)

Figure 8. Micrographs of laser damage on (a) Silicon, (b) diamond film on Silicon and (c) diamond film window.

the film may influence the damage threshold.

7. ACKNOWLEDGEMENT

This work was supported by NASA grant No. NAG-1-791.

8. REFERENCES

1. M. W. Geis, D. D. Rathman, D. J. Ehrlich, R. A. Murphy and W. T. Lindley, "High temperature point contact transistors and Schottky diodes formed on synthetic boron doped diamond", IEEE Electron Device Lett., EDL-8(8), 341-343 (1987).

2. J. F. Prins, "Bipolar transistor action in ion implanted diamond", Appl. Phys. Lett. 41(10), 950-952 (1982).

3. P. S. Panchhi and H. M. Van Driel, "Picosecond optoelectronic switching in insulating diamond", IEEE J. Quantum Electron. QE-22 (1), 101-107 (1986).

4. S. C. Rand and L. G. DeShazer, "Visible color-center laser in diamond", Optics Lett. 10(10), 481-483 (1985).

5. M. Sokolowski, A. Sokolowska, B. Gokieli, A. Michalski, A. Rusek and Z. Romanowski, "Reactive pulse plasma crystallization of diamond and diamond-like carbon", J. Crystal Growth, 47, 421-426(1979).

6. B. V. Spitsyn, L. L. Bouilov and B. V. Derjaguin, "Vapor growth of diamond on diamond and other surfaces", J. Crystal Growth, 52, 219-226(1981).

7. S. Matsumoto, Y. Sato, M. Tsutsumi and N. Setaka, "Growth of diamond particles from methane-hydrogen gas", J. Material Sci. 17, 3106-3112(1982).

8. A. Sawabe and T. Inuzuka, "Growth of diamond thin films by electron-assisted chemical vapor deposition and their characterization", Thin Solid Films, 137, 89-99(1986).

9. D. V. Fedoseev, V. P. Varnin and B. V. Deryagin, "Synthesis of diamond in its Thermodynamic Metastability Region", Russian Chem. Rev., 53, 435-444(1984).

10. Y. Hirose and Y. Terasawa, "Synthesis of diamond thin films by thermal CVD using organic compounds", Jap. J. Appl. Phys., 25(6), L519-L521(1986).

11. K. Suzuki, A. Sawabe, H. Yasuda and T. Inuzuka, "Growth of diamond thin films by dc plasma chemical vapor deposition", Appl. Phys. Lett. 50(12), 728-729(1987).

12. W. F. Krupke, M. D. Shinn, J. E. Marion, J. A. Caird and S. E. Stokowski, "Spectroscopic, optical and thermal properties of neodymium- and chromium-doped GSGG," J. Opt. Soc. Am. B, 3(1), 102-114 (1986).

13. R. M. Wood, Laser Damage in Optical Materials, Adam Hilger, Bristol (1986).

14. F. Forlani and N. Minnaja, "Thickness influence in breakdown phenomena of thin dielectric films", Phys. Stat. Sol. 4, 311-324(1964).

15. H. Goldenberg and J. C. Tranter, "Heat flow in an infinite medium heated by a sphere," Brit. J. Appl. Phys. 3, 296-298(1952).

16. T. W. Walker, A. Vaidyanathan, A. H. Guenther and P. Nielsen,"impurity breakdown model in thin films," in Laser Induced Damage in Optical Materials: 1982, H. E. et.al. ed. National Bureau of Standartds, 479-496(1984).

17. R. M. Walser, M. F. Becker and D. Y. Sheng, "Laser damage of crystalline silicon by multiple 1.06 -m picosecond pulses," in Laser Induced Damage in Optical Materials:1981, H. E. Bennet, et. al. ed., National Bureau of Standards Special Publication 638, 103-113 (1983).

18. D. H. Gill, B. E. Newman and J. McLeod, in National Bureau of Standard Special publication 509, 248(1977).

ADDENDUM

The following paper, which was scheduled to be presented at this conference and published in this proceedings, was cancelled.

969-18   **Optical transmission properties of crystalline diamond and diamond-like thin films in the wavelength region from 2 μm to 2 mm**
R. C. Carlson, J. C. Brasunas, D. E. Jennings, K. P. Stewart, J. B. Heaney,
NASA/Goddard Space Flight Ctr.

The following papers were presented at this conference, but the manuscripts supporting the oral presentation are not available.

969-04   **Ion-beam methods of producing diamond-like carbon films**
M. J. Mirtich, B. A. Banks, D. M. Swec, S. K. Rutledge, NASA/Lewis Research Ctr.

969-20   **Nucleation and growth of thin diamond films for tribological and optical applications**
K. V. Ravi, M. Peters, M. Pinneo, S. Yokota, L. S. Plano, Crystallume

969-23   **New carbon: silicon materials for optical applications**
S. C. Miller, Silicon Films Corp.; K. K. Mackamul, Utilities Power Group;
S. F. Pellicori, Pellicori Optical

AUTHOR INDEX